Sustainable Irrigation Management, Technologies and Policies III

WIT*PRESS*

WIT Press publishes leading books in Science and Technology.
Visit our website for the current list of titles.
www.witpress.com

WIT*eLibrary*

Home of the Transactions of the Wessex Institute.
Papers presented at Sustainable Irrigation III are archived in the WIT eLibrary in volume 134 of WIT Transactions on Ecology and the Environment (ISSN 1743-3541).
The WIT eLibrary provides the international scientific community with immediate and permanent access to individual papers presented at WIT conferences.
Visit the WIT eLibrary at www.witpress.com.

THIRD INTERNATIONAL CONFERENCE ON SUSTAINABLE IRRIGATION MANAGEMENT, TECHNOLOGIES AND POLICIES

Sustainable Irrigation III

Organised by

Wessex Institute of Technology, UK
University Politehnica of Bucharest, Romania

Sponsored by

EurAgEng: European Society of Agricultural Engineers
CIGR: International Commission of Agricultural Engineering
WIT Transactions on Ecology and the Environment

WIT Transactions

Transactions Editor

Editorial Board

Sustainable Irrigation Management, Technologies and Policies III

Editors

C.A. Brebbia
Wessex Institute of Technology, UK

A.M. Marinov
University Politehnica of Bucharest, Romania

H. Bjornlund
University of South Australia and *University of Lethbridge, Canada*

WIT*PRESS* Southampton, Boston

C.A. Brebbia
Wessex Institute of Technology, UK

A.M. Marinov
University Politehnica of Bucharest, Romania

H. Bjornlund
University of South Australia and University of Lethbridge, Canada

Published by

WIT Press
Ashurst Lodge, Ashurst, Southampton, SO40 7AA, UK
Tel: 44 (0) 238 029 3223; Fax: 44 (0) 238 029 2853
E-Mail: witpress@witpress.com
http://www.witpress.com

For USA, Canada and Mexico

Computational Mechanics Inc
25 Bridge Street, Billerica, MA 01821, USA
Tel: 978 667 5841; Fax: 978 667 7582
E-Mail: infousa@witpress.com
http://www.witpress.com

British Library Cataloguing-in-Publication Data

A Catalogue record for this book is available
from the British Library

ISBN: 978-1-84564-446-8
ISSN: 1746-448X (print)
ISSN:1743-3541 (online)

The texts of the papers in this volume were set individually by the authors or under their supervision. Only minor corrections to the text may have been carried out by the publisher.

Printed in Great Britain by Martins the Printers.

Preface

This book contains most of the papers presented at the Third International Conference on Sustainable Irrigation, Management, Technologies and Policies held in Bucharest and organized by the University Politehnica of Bucharest and the Wessex Institute of Technology of the UK with the collaboration of the European Society of Agricultural Engineers and the International Commission of Agricultural Engineering.

The meeting followed the success of the first two Conferences which were organised in Bologna (2006) and Alicante (2008). The meetings discuss ways to achieve the most efficient and equitable use of water resources applied in irrigation, at the same time as ensuring their sustainability.

The objective of the Conference Series is to provide a venue for an international presentation and dissemination of papers related to sustainable ways of conducting irrigation. Sustainable irrigation has three key aspects, i.e. i) developing ways of using scarce water resources most efficiently to maximize food production while minimizing the environmental impact of irrigation activities on the affected land and water resources; ii) ensuring that the extraction of water from surface and groundwater sources is set at sustainable levels to maintain healthy aquatic ecosystems to secure a continued supply of quality water for irrigation, recreation and urban/industrial uses; iii) facilitate an equitable distribution of the water available for consumption between competing users, which is essential to maintain sustainable communities.

The papers presented in this volume have been classified in the following sections: Irrigation Management; Modelling and Simulation; Farmer's Actions, Environmental Benefits and Socio-Economic Consequences.

The Editors are indebted to Professor Ecaterina Andronescu, Rector of the University Politehnica of Bucharest, for her support of the meeting and having accepted to act as Honorary Chairperson. They are also grateful to all authors for the excellent standard of papers and the members of the International Scientific Advisory Committee and other colleagues who helped to select contributions, ensuring in this manner the quality of the book.

The Editors, 2010

Contents

Section 2: Modelling and simulation

Section 3: Sustainable irrigation – farmers' actions, environmental benefits and socio-economic consequences Special session chaired by H. Bjornlund

Section 1
Irrigation management

Land reclamation and water management in an arid region: a case study of Xayar county in Xinjiang Uygur autonomous region, China

T. Yamamoto[1], A. Jalaldin[2], A. Maimaidi[3] & T. Nagasawa[1]
[1]*Research Faculty of Agriculture, Hokkaido University, Japan*
[2]*College of Resources and Environmental Sciences, Xinjiang University, China*
[3]*Xayar Water Resource Agency, China*

Abstract

Irrigation is indispensable in guaranteeing agricultural production in arid regions, such as the Tarim Basin. In this region, administrations and farmers have reclaimed wild land, and have attempted agricultural development by securing the water resources. However, this area faces problems with an extremely irrational plan between land reclamation and water resource development, which has resulted in an acute shortage of water in this region. The situation has also caused ecological damage. This paper discusses this region's problems with agricultural water use and farmer's activities. It is shown that water supply volume per unit of farmland is tending to decrease. It is also clear that the water supply from the river has decreased greatly in the region where water-saving irrigation by groundwater pumping was introduced earlier. Water-saving irrigation is being introduced to resolve the water shortage and create new water resources. Although it seems to be effective for saving water resources, improper management hinders effective irrigation in this region. In particular, considering that water management is different among landowners and farmers, it has become clear that there are several problems – lack of an adequate management system and training system for facilities maintenance, etc. – that need to be resolved for water-saving irrigation in the region.

Keywords: Tarim River, water-saving irrigation, land and water management.

doi:10.2495/SI100011

1 Introduction

In arid regions, agriculture and human activities are possible only where there is a source of water. Desert oases exist because of the water resources located there. The Taklamakan Desert is located in Central Asia, in the Xinjiang Uyghur Autonomous Region of the People's Republic of China. It is bounded by the Kunlun Mountains to the south, and the Pamir Mountains and Tian Shan to the west and north. A large area of farmland exists in the border area of the Taklamakan Desert, an achievement that was made possible through the development of water resources and the construction of irrigation facilities. Expansion of food demand and industrial output through population increases in China are major factors in the development of agriculture in such regions where water resources are inadequate. Development of farmland and water resources has been facilitated through an injection of both capital and technology involving (1) the Xinjiang Production and Construction Corps (a unique economic and semi-military governmental organization existing in the Xinjiang Uyghur Autonomous Region) and (2) the Great Western Development Strategy, a governmental policy aimed at opening up less developed regions of China. These projects are geared toward agricultural promotion through the reclamation of farmland from deserts and the development of water resources. However, water shortages have occurred because of irrational planning of land reclamation and water resource development [1]. In addition to water shortages for agricultural production, the regional and river basin ecosystem has become degraded due to human activity [2]. In attempting to combat water shortages in this region, the effective use of groundwater, which exists in large amounts, and the reduction in water loss through lining the canals are both effective measures [3]. On the other hand, because groundwater levels depend on the extent of seepage of irrigation water [4], the groundwater resources decrease if seepage is prevented. Moreover, exhaustion of groundwater resources by widespread pumping has been recorded by Xinan Deng et al. [5] and others. To summarize, sustainable and effective utilization of groundwater is not a simple matter.

In this study, problems relating to agricultural water use are discussed against a background of actual conditions of water resources and agricultural activities under irrigation, especially water-saving irrigation, in Xayar County, Tarim River basin. Future water management and effective water resources are also examined.

2 Methods and outline of investigation area

The investigation targeted Xayar County, which is located in the Taklimakan Desert in the northern border region of Xinjiang Uyghur Autonomous Region. Xayar County is within the Aksu Prefecture and is located in downstream of the Ugen River, which is a branch of the Tarim River (Fig.1). The county population is 174,000 and its area is 31.9×10^2 km^2. The main local crop is cotton, and irrigation water relies on the Kyzyl Dam upstream of the Ugen River. Many

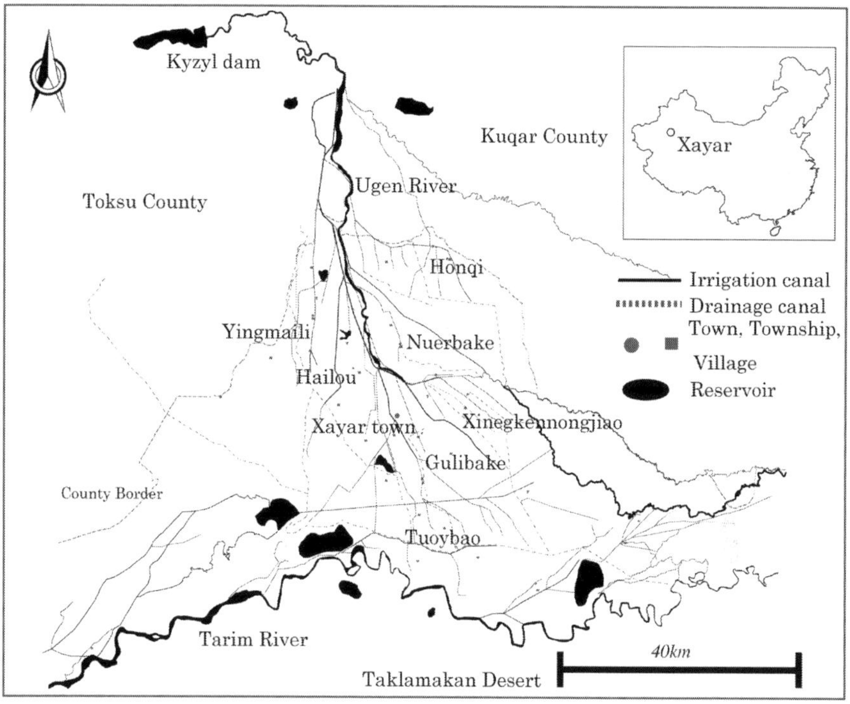

Figure 1: Investigation area.

irrigation resources in the Ugen River basin are derived from a mountain glacier. The Kyzyl Dam supplies water to three counties: Xayar, Kuqar, and Toksu. The administrative and irrigation districts are numerous. The Xayar County administration is divided into eight "townships and towns" (from upriver down; Honqi, Yingmaili, Nuerbake, Hailou, Xayar Town, Xinkennongjiao, Gulibake, and Tuoybao) in the Xayar irrigation district. Irrigation water from the Ugen River is supplied to these townships and towns. However, part of the intake is from the Tarim River in Tuoybao Township, and there are several family-run farms in this district. On the other hand, in Xinkennongjiao, the farmland has been developed through national enterprise, with many landowners and a few family-run farms.

To ascertain the prevailing state of irrigated agricultural enterprise, interviews with representatives of branch offices of the local water resource agency and some farmers were conducted from 2003 to 2009. In the interviews, we mainly researched the irrigation method, irrigation frequency, problems in irrigation, facilities management, farmers' income, and family structure, etc. Data concerning the irrigation water supply from 1998 to 2008 and for farmland areas in townships in Xayar County in 2007 were obtained from this agency. The water supply data is the daily volume of irrigation water distributed from the main irrigation canal to each township and town.

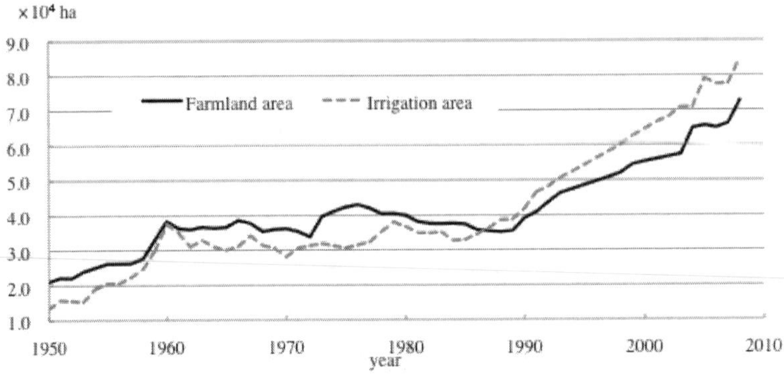

Figure 2: Changes in cultivated and irrigation areas in Xayar County. This figure was created by modifying data from Jalaldin et al. [1].

3 Farmland development and land-use rights

3.1 Changes in farmland and irrigation areas

Figure 2 shows changes in the farmland and irrigation areas of Xayar County since 1950. Since that year, farmland areas have greatly increased under China's Great Leap Forward policy, but the expansion rate eventually slowed. An increase was again seen, fuelled first by the inflow of capital and technology under the Economic Reform policy of the 1990s and then through the Great Western Development Strategy starting in 2000. The current agricultural area is now three times greater than that in 1950 (i.e., 60,000 ha). Although the irrigated area was smaller than the farmed area until 1987, the former is now increasing each year at a greater pace than the latter. The main factors leading to the increase in irrigation are the advances made in irrigation systems through the investment of capital and technology and the consequent potential for double cropping. The Kyzyl Dam, which was completed in 1991, has facilitated a far greater reliability of irrigated water supply, circumventing the need to obtain water from the Ugen River.

3.2 Family-run and landowner farmers

Farm ownership affects every agricultural activity (Fig.3). From results of the interview research, we considered the effects of farm ownership. In China, ownership of all farmlands belongs to the nation. This began during the People's Commune period. After the Economic Reform Policy, only the 30-year's land-use rights were given for the purpose of expanding farmers' independent management to increase food production. In other words, the current farmers borrow land-use rights through contracts with the nation. First, farmland (i.e., 30 year's contract farmland) is distributed equally among farmers in a village, and

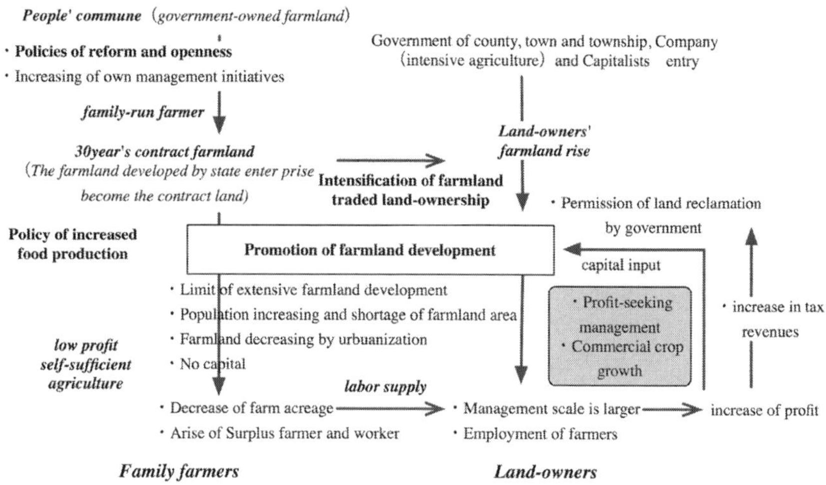

Figure 3: Concept of ownership of the land-use rights.

each farmer owns land-use rights. But the land-use rights in newly developed farmlands are sold because specific farmers cannot own these rights, and the original farmers do not stay on the farms. At this time, people have become landowners because of taking land-use rights through administrative organizations, companies, and capitalists. Many landowners pushed forward scale expansion of the management to buy not only the rights of developed farmland but also general farmers' farmland. Farmland developed by not only state enterprise but also each level of administration organization or individuals (rich farmers, companies, capitalists) under the food production increase policy. It was difficult to expand the farmland area in some old village that has little development appropriate place. In addition, when the population increased, the management scale per a farmer decreased because the farmland need to be divided into new family and farmers are limited to leave from the village in China. So the original farmers seemed hardly to management with advantage of scale. As a result, small-sized agricultural management on self-support crops was mainly in many farmers. However, the farmers who owned the farmland under optimum production conditions or who have the opportunity for scale expansion can get some profit to cultivate commercial crops (cotton and fruits) with self-supporting crops. The landowner farmers have enabled the large-scale management of commercial crops to use the surplus labour of those that have too little farmland to be self-sufficient or have no land-use rights. In addition, they have further expanded and construct the new irrigation facilities. Although new land reclamation is currently subjected to rigid control, the permission of land reclamation for landowners tends to be accepted easily because of the increase in tax income incurred.

4 Water management and changes in supplied water

4.1 Irrigation methods and maintenance rules

Many irrigation resources in the Tarim River basin are derived from a mountain glacier, therefore, the volume of irrigation water increases in the summer but decreases in the winter. Although the Kyzyl Dam constitutes the main water resource in this region, there is a very distinct seasonal bias, with inflow and outflow for irrigation being concentrated in the summer months [6].

The irrigation water is supplied by the Longkou diversion headworks to the Xayar Main Irrigation Canal (length, 50 km). The water is delivered using a five-step canal irrigation system [6]. In this county, the lining rate over the total canal length is only 4%, conveyance efficiency is about 60%, and irrigation efficiency is only 40%. Irrigation is performed mainly from the spring through the summer, and minimal in winter. From the spring through the summer, irrigation water is used for general crop growth. Winter irrigation aims at maintaining soil moisture, which would otherwise be low in the early spring sowing period. Border and furrow irrigation occupy about 80% of farmland in Xinjiang [7], and it is similar condition in this region. In addition, water-saving irrigation technology has been introduced in recent years.

The water resource agency in this county is responsible for maintenance of the main canal and water distribution to each township or town, and each branch office is responsible for the management of water distribution and maintenance within each settlement. Water supply to each farm is overseen by local office staff and farmers' representatives. When the irrigation water supply is adequate, the appropriate level of water is distributed for each crop and planted area. However, during shortages, irrigation for all the fields is not possible. Therefore, the following rules exist when establishing priority [8].

Spring–summer:

• Upriver takes priority over downriver.

• Larger canals (or irrigation blocks) take priority over smaller canals.

Winter:

• Downriver takes priority over upriver, because water shortages are more likely downstream at sowing time.

• Wheat fields and land under 30 years' contract take priority.

4.2 Monthly irrigation water volume

Figure 4 shows the variations in the monthly supply of water per unit area to each town and township from 1998 to 2008. The irrigation area values in each town or township were not obtained except in 2004 and 2007, and these values are presumed from two years of data. The largest volume is supplied in July, followed by August, partly because of high summer temperatures. March and November also show peaks: the former being when sowing occurs and the latter when irrigation is required for wheat. Although irrigation is also required in December and February, this is winter irrigation, as explained above.

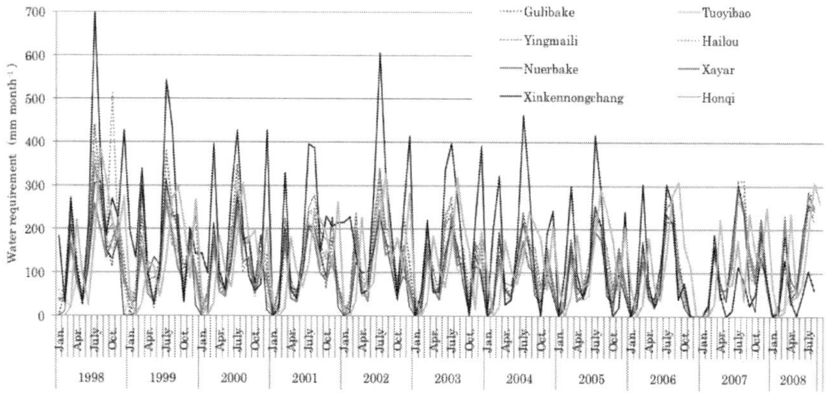

Figure 4: The irrigation water supplied in each township and town.

The water supply is found to decrease in each town or township. The decrease was small for all towns and townships except for Xinkennongjiao Township. This reduction in irrigated water supplied is remarkable in Xinkennongjiao Township. One of the reasons for such a decrease may be the abandonment of farmland. However, we cannot detect any reason that farmland decreased. Therefore, the reduction in irrigation water supplied is considered to be attributable to other factors. Maintenance levels of the irrigation canal have not achieved a higher rate of conveyance efficiency in this particular township nor has there been any major shift in the types of crops grown locally. It is therefore believed that irrigation water must be supplied from other resources, and it is surmised that an increase in groundwater is the reason. According to the authorities, since 2004 a large area in Xinkennongjiao Township has been able to reduce requirements for irrigated water by using groundwater. Why has this not occurred in any other township? The main reasons are that the economic burden involved in construction of facilities is large, and there is inadequate technical input for water-saving irrigation.

4.3 Water fees and water-saving consciousness

We considered the influence on water-saving consciousness by water fees from results of interview to Water Resources Agency and its brunch offices. Before 2000, the water fee per unit area of crops was fixed, and farmers paid taxes according to their own farmland areas. Therefore, because the payments did not change no matter how much irrigation water farmers used, farmers were not motivated to save water. In addition, many complaints occurred in the downstream farmland located far away from the water resource, because enough irrigation water did not reach to waste in the upper farmlands. The government decided that new water price rules should depended on the irrigation water volume in a wider area of this Autonomous Region after 2000. As a result, although wasting of irrigation water and feelings that water fees are unfair have declined, the water shortage has not yet completely resolved. In addition, the uniform maintenance level of irrigation facilities is demanded to apply new water price rule. However, actually, there is also an area where the maintenance level is insufficient (lining rate is too small), and the introduction of the new rule is difficult. So it is in situation that an old rule and a new rule are being used appropriately according to the maintenance level of the irrigation facilities. In many farmlands, the cost of water is collected under such a rule. On the other hand, when water-saving irrigation facilities are improved individually (in many cases the irrigation water depends on groundwater), it doesn't require the water fee of groundwater basically. Therefore, the incentive for saving water costs occurs through advancing water-saving irrigation. But because electricity costs are needed for pumping up groundwater, many farmers believe that the cost of electricity is higher than the cost of water or the facilities cost is higher.

5 Problems with introducing water-saving irrigation

5.1 Introduction of water-saving irrigation

Chronic water shortages are a problem in this region, and the introduction of water-saving irrigation has been proposed as a solution. The following matters were found in the interview researches. Since 2007 a water-saving project has been in place for farmers in Xayar County. Water-saving irrigation facilities (i.e., drip irrigation) have been improved for about 10,000 ha (150,000 mu in Chinese units) of farmland, equal to about 10% of the total cultivated area in the county. The Chinese government subsidized one-third of the total cost of construction. The remaining cost is borne by the county and farmers. Farmers paid roughly 800 CNY per 0.067 ha (about 1 mu). However, irrigation facilities cannot necessarily be constructed on any farmland. Local construction standards recommend one pumping facility for irrigation of about 27 ha (400 mu). Construction of this facility is difficult for the common family farmer because each farmer in this region manages about 1.3 to 2.0 ha (20–30 mu). Although this is about 6.7 ha in Tarim Township, the construction of a pumping facility is not cost-efficient for the one family farmer and it is never a cheap investment, considering that their profit from production is about 1,200 to 2,000 CNY per year. On the other hand, some farmers indicated that the irrigation volume by river intake decreased to one-third of that before using water-saving irrigation. Some farmers in Xinkennongjiao Township and the water resource agency branch office staff in Tarim Township stated that cotton production increased as result of water-saving irrigation.

5.2 Problems with management of water-saving irrigation facilities

Limiting construction of these facilities by farmland scale does not arise for the landowners because their land holdings are large enough to justify the construction of facilities on their land. The landowners (or the responsible farmers employed) manage these facilities by themselves. On the other hand, in the case of the common farmers, their representatives control each pump. In addition, even if irrigation priorities are decided, some farmers cannot understand correct operation of the facilities provided, or there may be a lack of cooperation with other farmers. Therefore, each farmer irrigates independently; as a result, the water pressure in some irrigation pipes decreases, causing shortages of irrigation water. Naturally, the management of a pump is unified, and the rotation rule of which farmland irrigate is not also defined correctly. On occasions, disruption in the water supply occurs because the pipes upriver are cut to monopolize supplies. In the case of some common farmers in Honqi Township, the water-saving irrigation was costlier than irrigation from river intake because water-saving irrigation could not be managed effectively. Problems are exacerbated because farmers are only given a manual to learn about the scheme, and the manual is in Chinese, which may not be their spoken language. The water resource agency can't respond to this problem because of its insufficient budget.

6 Assignment for sustainable irrigation

6.1 A cause of water shortage

Irrigation water demands (design net water requirement) and requirements in Xayar County are shown in Figure 5 representing design net water requirements, water requirements (supplied irrigation water), and presumed net water requirements (water requirement × 0.4: irrigation efficiency in this region [1]), respectively. The total design net water requirement is calculated from the cultivated area according to crops grown in the years 2004 and 2008 and to the design net water requirement for each crop. Data for each cultivated area from 2005 to 2007 that were not available were calculated from the transition in cultivated field area over the same period. C is less than A between 2004 and 2008, and the average annual shortfall volume is 7.6×10^7 m^3. It is thought that there is sufficient water requirement (B), although many farmers argue strongly that there is a shortage of water.

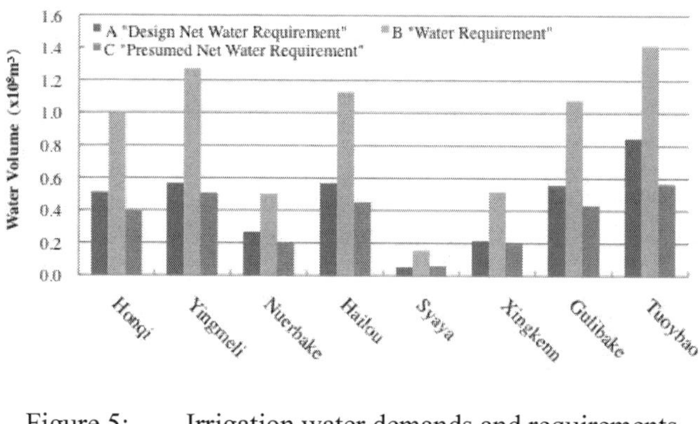

Figure 5: Irrigation water demands and requirements.

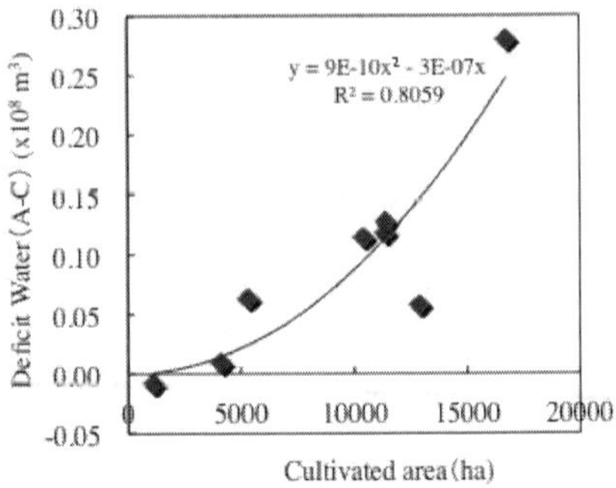

Figure 6: Relationship between deficit water and cultivated area.

The relationship between water volume deficit (B – C) and the cultivated field area shows that the former becomes larger by expansion of the cultivated area (Fig. 6). Because expansion of the cultivated area indirectly indicates the total length of the main channel, it is suggested that a water shortage has occurred because of seepage from both canal and irrigation channels.

6.2 Relationship of land reclamation to water resource development

We think that an irrational relationship between farmland and water resources developments is one of the factors causing water shortage in this region. In Japan, farmland reclamation and water resources development are usually planned integrally. Developed farmland area limits the necessary water resource scale, or an irrigation area is decided with the amount of water resources that it can supply. However, in this region, the section that permits the farmland reclamation and the section that take charge of water resource development and management were separate. As result, the adjustment in the development plan (also after development) is insufficient, making water supply and demand unbalanced.

The section that permits farmland development is different from a prefecture, a county, and a township, depending on the size of reclamation area, and the influence of the water management section (i.e., water resource agency) on this permission is barely reflected. But the section that manages and supplies water is responsible for distributing irrigation water equally to the existing farmland. Therefore, a shortage of water develops naturally if farmland extends without water resources developing. At this time, there is administrative instruction by the Autonomous Region or Prefecture governments; however, the effect is weak because there are uncertainties and insufficient statistics data to understand irrigation area [2]. It is also considered the section that permits farmland development judged there was enough area of possible reclamation area, because they didn't understand certain data. As result, they permitted the new farmland development over the limit of water supplied. Such a vicious circle has aggravated the water shortage.

6.3 Requirements of effective utilization of water resources

It was indicated that the water shortage has occurred in each field due to water seeping through the canal; however, the volume of irrigation water supplied was sufficiently in Xayar County (Fig. 5). This problem may be solved by improving the efficiency of the water supply by lining the canal. However, the cost of lining the entire canal is immeasurable. In addition, it can be predicted that it will be necessary for long-term for restoration of canals' freeze-fracturing. Therefore, to secure irrigation water in a short time period, the introduction of saving-water facilities is regarded as an effective method and the administrative organization does not need to manage and construct for individual facilities. Some farmers said in their interviews that the saving-water facilities are effective in conserving water resources and increasing production. Such an effect is confirmed under the appropriate management, and the general farmers have a problem on the management. It is necessary to support these farmers to achieve the certain effects of water-saving irrigation. For this purpose, important factors are (1) financial support more than current, (2) the establishment of a training system for facilities maintenance, and (3) substantially raising the farmers' water-saving consciousness.

7 Conclusion

This study has shown that water-saving irrigation is effective in saving the water resources from the daily supplied water data and the interview researches. However, the introduction of water-saving irrigation is not necessarily effective in all cases. It is evident that the effects of water-saving irrigation are different for the landowners versus common farmer, even if facilities are similarly constructed. The differences in land-use rights and the amount of capital are large problems in the introduction of water-saving irrigation.

It was also confirmed that the main cause of water shortage is seepage from branch canals before reaching farmlands; the volume of supplied water is not the problem.

The development of water-saving irrigation will be indispensable to the preservation of water resources in this region in the future. The influence of pumping of groundwater on natural vegetation is of concern, and the environmental impact of this on large areas has not yet been realized. It will be necessary to examine these environmental concerns, including evaluating groundwater resources.

Acknowledgements

Research was supported by Xayar County Government, Xayar Water Resource Agency, Xinjiang University and Xinjiang Uyghur Autonomous Region Water Resource Agency.

References

[1] Abdisalam Jalaldin, Nagasawa Tetuaki & Yamamo Tadao (2005). Irrigated agriculture and regional environment in the Tarim River watershed – case study of the shaya irrigated district in the Ugen River basin -. *Journal of Rural Planning* 7, 73-78.

[2] Q. Feng, K.N. Endo & G.D. Cheng (2004). Towards sustainable development of environmentally degraded arid rivers of China – a case study from Tarim River –. *Environmental Geology* 41, 229-238.

[3] Fan Zili, Xia Xuncheng, Shen Yuling, Alishir Kurban, Wang Ranghui, Li Shemgyu & Ma Yingjun (2002). Utilization of water resource, ecological balance & land desertification in the Tarim Basin, Xinjiang. *Science in China*, 45, 102-108.

[4] Ma Jin-zhu, Li Ji-jun (2001). The groundwater resources and its sustainable development in the south edge of Tarim Basin. *Chinese Geographical Science 11(1)*, 57-62.

[5] Xinan Deng, Yuzhou Luo, Sucheng Dong & Xiusheng Yang (2005). Impact of resources and technology on farm production in northern China. *Agricultural Systems* 84, 155-169.

[6] T. Yamamoto, T. Nagasawa & A. Jalaldin (2006). Irrigated agriculture and the environment in the Tarim River watershed: case study of the shaya irrigation district in the Ugen River basin. Sustainable Irrigation Management, *Technologies and Policies*, WIT Press, 303-312.

[7] Zhong Ruisen, Dong Xingyang & Ma Yingjie (2009). Sustainable water saving: New concept of modern agricultural water saving, starting from development of Xinjiang's agricultural irrigation over the last 50 years. *Irrigation and Drainage* 58,383-392.

[8] Yamamoto Tadao, Abdisalam Jalaldin, Anwaire Maimaidi & Nagasawa Tetuaki (2010). The present condition and problem of agricultural water management in northern part of Taklamakan Desert. *1st International Conference on Environmental and Rural Development* (in press).

Making large gravity irrigation systems perform

G. Spencer
Business Development Manager, International, Rubicon Systems, Australia

Abstract

By 2050 the world's population will increase by an estimated 50%. Most of the increased food demands will have to come from productivity improvements in existing irrigated agricultural areas. These improvements will need to be achieved with less water and less farm land because of the competition from urbanization, industrialization, the need to provide water for environmental purposes and to respond to climate change impacts. Therefore, the challenge facing us is to make existing irrigation systems more efficient and productive to better utilize the shrinking agricultural land and water resources.

In south-eastern Australia, severe water shortages brought about by a prolonged 11 year drought sequence has spurred a lot of innovative thinking about how to use the limited available water more efficiently in irrigation and, at the same time, improving the performance of irrigation systems.

A higher level of technological investment in the modernization of large unlined, gravity fed irrigation systems in the south-eastern state of Victoria has resulted in increases in efficiency from about 70% up to about 90% - a remarkable outcome. In Victoria, the water saved is being reallocated equally between urban and industrial users, the environment and to existing farmers to improve their security of supply.

The investment required is a mix of both hardware and software involving remotely controlled gates, accurate water measuring devices, Supervisory Control and Data Acquisition (SCADA) and system management software. This investment, combined with modelling and control technologies, has resulted in the delivery of significantly improved services to irrigators in terms of equity, reliability and flexibility, and huge improvements in the management capability and performance of existing large-scale irrigation systems.

WIT Transactions on Ecology and the Environment, Vol 134, © 2010 WIT Press
www.witpress.com, ISSN 1743-3541 (on-line)
doi:10.2495/SI100021

Keywords: modernization, SCADA, channel control technology, accurate water measurement, improved performance and efficiency, water resource savings.

1 Introduction

South eastern Australia is in the grip of an unprecedented 11 year drought sequence, which has significantly reduced the availability of water in the large inter-state Murray-Darling River basin system. Annual River flows are at all time low levels and since 1997 have averaged 44% less than the long term average (1891-1997) [2]. The state (provincial) government of Victoria (GoV) is facing a challenging and constantly evolving situation of: (i) increased water demands from a growing population; (ii) continued reliance on reliable water supplies to meet agricultural, industrial, tourism and recreational demands; (iii) the need to provide water for environmental purposes to the increasingly stressed natural environment; (iv) the continuing drought sequence, now in its twelfth year; and (v) climate change predictions of even less rain and higher temperatures. Faced with this situation, the government has formulated and is implementing a state water plan entitled "Securing Our Water Future Together" [2] with the over-arching objective of improving the use and management of the state's diminishing water resources on a sustainable basis.

The six priority elements of the state water plan are: (i) improved water allocation framework; (ii) restoring rivers and aquifers to sustainable levels; (iii) smarter use of irrigation water; (iv) smarter use of urban and industrial water; (v) water pricing for sustainability; and (vi) innovative and accountable management of the water sector. This paper details the GoV's third priority area - the smarter use of irrigation water [2].

2 Improving the efficiency and productivity of irrigation

Like anywhere in the world, irrigation in Victoria generates substantial and growing economic and regional benefits and uses the largest share of available water. Water consumption in all sectors state-wide is about 7 Billion cubic metres (BCM) per annum, with the irrigation sector using about 77% of this figure. Figure 1 shows the consumptive uses of water by sectors (1996-1997).

Compared to other sectors, irrigation uses its large share of water relatively inefficiently. The key challenges for the GoV to bring about the smarter use of water for irrigation are to: (i) ensure that the irrigation sector will prosper in a competitive, global environment by moving water to higher-value uses through the enhancement of water trading and water markets; (ii) reconfigure irrigation infrastructure and adopt world best practice innovations and new technology through large scale capital investment in modernization programs; and (iii) adjust to changing community expectations about the environment by improving on-farm water use, reducing losses in distribution systems and allocating water to the environment.

The outcomes expected are: (i) a confident, profitable and adaptable irrigation sector that generates wealth for regional communities and Victoria;

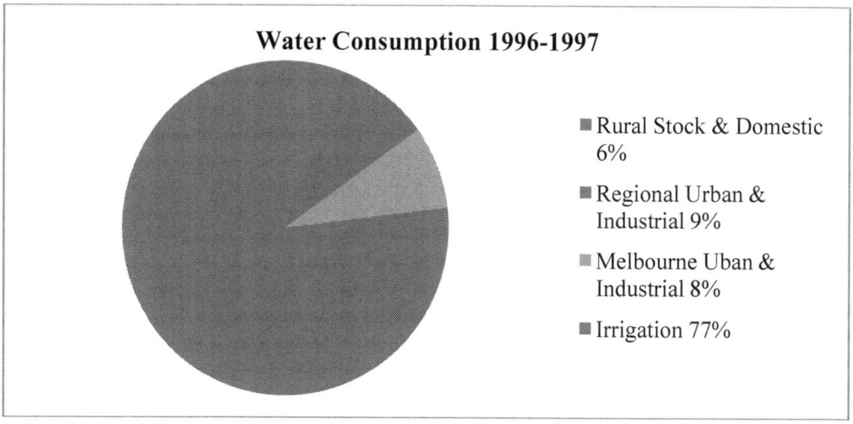

Figure 1: Water consumption in Victoria by major consumer group [3].

(ii) an increase in the efficiency of irrigation systems across the state of 25% by 2020; (iii) responsive and flexible irrigation supply systems that are effective and economic; (iv) lower salinity and nutrient side-effects of irrigation; and (v) the freeing up water to better meet future urban, industrial environmental needs.

Many of these expected outcomes will be delivered through a major government initiative and investment program, the Northern Victoria Irrigation Renewals Project (NVIRP), that will bring about increased efficiency of use of irrigation water and greatly improved irrigation water services to customers in a far more responsive, effective and economical manner.

3 The Northern Victoria Irrigation Renewals Project (NVIRP)

The NVIRP [4] is an AU$2bn, 2 stage program with the primary objective of improving water distribution efficiencies in order to save a total of 425 MCM/annum of irrigation water (225 MCM/annum in stage 1 and a further 200 MCM/annum in stage 2). The stage 1 savings will be shared equally between Melbourne urban and industrial users, the environment and irrigators. Stage 2 water savings will be shared equally between the environment and irrigators. In addition, the project aims to greatly improve levels of service to farmers and so improve the productivity of irrigation.

The project area (stages 1 and 2) covers about 800,000 ha, involving about 14,000 farms and 6,300 km of publically owned channels, fig. 2.

3.1 Unaccounted for water and improved operating efficiency

Water unaccounted for in open, unlined irrigation systems consist of evaporation, inaccurate metering and outfalls. Outfalls are generally unplanned spills or releases of water that prevent water over-topping of channel banks as a result of poor regulation and control of water flows. Other sources are leakage, seepage and losses from natural water courses (rivers and creeks) that are used as part of the water delivery network. The typical percentage of unaccounted for water from each source for the irrigation systems within the NVIRP project area are shown in figure 3.

The typical operating efficiency, from reservoir to farm supply point is about 70%. The NVIRP water savings are generated from a reduction in some of the losses; particularly those associated with inaccurate measurement within irrigation systems and at each farm supply point; better water regulation to

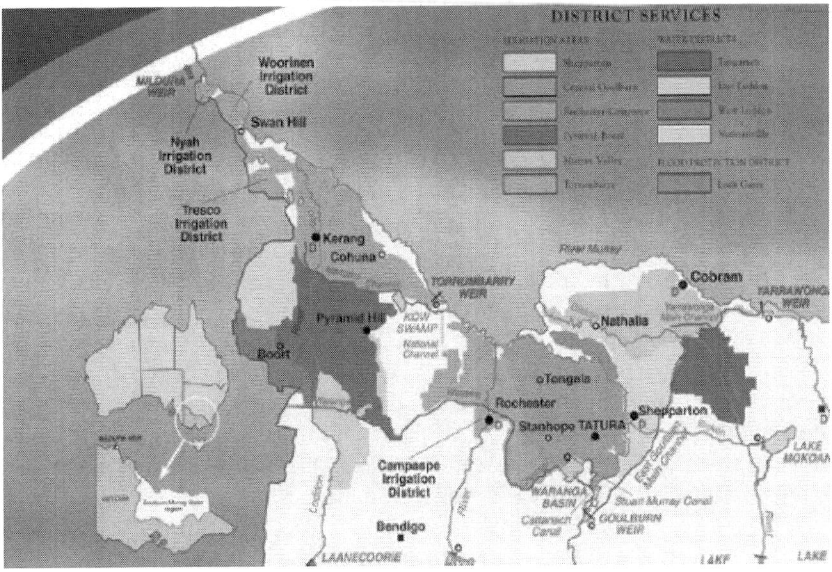

Figure 2: NVIRP project area [4].

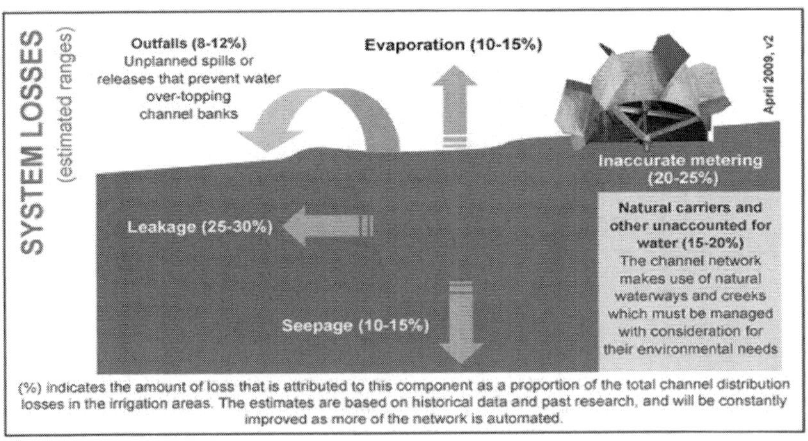

Figure 3: Typical unaccounted for water in irrigation systems [4].

Figure 4: Sources of channel losses. L-R, inaccurate measurement at farm supply point; old timber drop-bar weir type regulator and measuring point manually operated; operational losses resulting in "outfalls".

virtually eliminate operational losses from outfalls; and leakage and seepage through selected lining of channels.

The NVIRP investment program is funding activities that will deliver a state-of-the-art modernized irrigation system and completely transform the existing manually operated, labour intensive, conservatively operated, high water loss system into one that is extraordinarily responsive, flexible, equitable and efficient, enabling irrigators to get water at the flow rate and time they specify. The expected outputs from the investment will be: (i) channel automation with full SCADA capability; (ii) accurate measurement of water at all levels; (iii) reduced leakage and seepage through selected channel lining; (iv) reduced operational losses through enhanced management; and (v) significantly improved water services to customers.

3.2 Total Channel Control® (TCC®) technology

To deliver the expected outcomes, NVIRP is procuring Rubicon Systems Australia's Total Channel Control (TCC®) [5], which is a revolutionary end-to-end irrigation canal automation technology that is transforming the inefficient manually operated open canal networks into fully automated, integrated and remotely controlled systems that are achieving demonstrated new benchmark delivery efficiencies of up to 90% [6].

A breakthrough in both irrigation control and flow measurement, the TCC® technology provides a system of solar-powered channel regulators and gates, linked through radio telemetry and advanced computer software to enable the whole channel network to operate efficiently, automatically and remotely. Solar power at each gate eliminates the cost of providing mains power to each site. This is particularly important when sites are in remote locations away from the power grid.

3.2.1 The Rubicon FlumeGate™

A fundamental component of TCC® is Rubicon's unique FlumeGate™ (fig. 5), which provides accurate measurement and control/regulation of flows at every point throughout an entire system, while eliminating manual operation of all water control gates.

Water is measured accurately through each gate for both free over-fall and submerged flow conditions, using the unique flow measurement technology developed by Rubicon and integrated into the design of each gate. Combined with control algorithms and the high level of measurement accuracy at each structure, leakage, seepage and theft losses can be identified and pinpointed more accurately. The comprehensive management information generated by TCC® is then used as the basis for selective channel lining investment by NVIRP and for routine maintenance and follow up action by the Irrigation Management Authority (IMA). For NVIRP, this improves the efficiency of investment considerably in that it enables just the problem reaches of channels where the greatest losses are occurring to be addressed. Channel lining activities typically involve laying flexible high density polyethylene (HDPE) membrane, 2mm thick (fig. 6). Lining involves re-shaping the existing earthen channel before the lining is placed and fencing of channel rights-of-way to exclude cattle from damaging the lining.

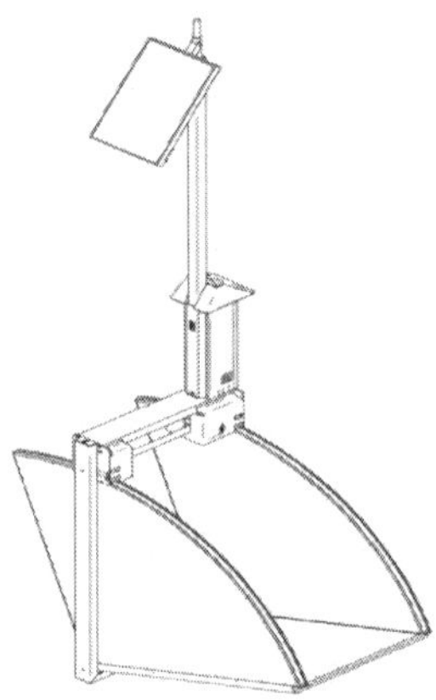

Figure 5: Schematic diagram of a Rubicon FlumeGate ™ complete with communications pole, solar panel, electronics module, gear box and drive shaft.

Figure 6: Typical NVIRP channel lining activities.

Lastly, automatic control of open channels and flow measurement is realizing significant water savings that are being reallocated to urban users (who are funding a significant portion of the project); to distressed river networks; and, to farmers to improve their security of supply. Furthermore, the prevention of outfalls ensures that pollutants such as salt and nutrients do not enter river systems and cause environmental damage.

3.2.2 Installation of FlumeGates™

An advantage of the FlumeGate™ is the quick installation into existing regulation structures where the concrete is sound, resulting in considerable cost savings through not having to replace the whole structure. Installation can be done in "wet" or "dry" conditions, resulting in minimum disruption to irrigation supplies. Figure 7 below shows gate installations under varying conditions. The right-hand photo shows details of the frame that is attached to the existing structure that then accepts the FlumeGate™ module. Where a new structure is required, a low cost alternative has been developed using a pre-cast concrete components.

3.2.2.1 Calibration and commissioning Each FlumeGate™ is ready to "plug and play" as it is calibrated and pre-commissioned at the factory and comes "SCADA ready" with factory set configurations. Each gate has dual redundancy water level sensors incorporated and connected to high precision digital instruments. There is no requirement for commissioning in the field. All that is required is simple accuracy verification and field calibration with no specialist skills required by field staff. Figure 8 shows FlumeGates™ installed and functional.

4 Enhanced irrigation management and improved customer services benefits

While water savings is the primary focus for NVIRP stage 1, the modernized irrigation system is delivering significant new benefits for irrigators and IMA

Figure 7: FlumeGate ™ installation in "wet and dry" channel conditions.

Figure 8: (L-R) New 5 bay regulator and a major regulating and measuring point retro-fitted to existing concrete structures. Note on the right the reduced number of bays required due to enhanced flow characteristics.

water operational managers and staff alike. The investment in stage 1 activities is also laying the foundation for substantial on-farm water savings, which will be the focus of investment under stage 2.

Modernisation of the irrigation delivery system is the first step in revolutionising how rural water is valued and managed. When completed, the modernized system will provide improved levels of service to customers and a more efficient network for system operators to manage. These benefits are evident from performance data from the 2008/09 irrigation season in areas already modernized. Results, using data from the much enhanced management information systems (MIS) software provided as part of the TCC® package, clearly show significant improvements in irrigation system performance in key areas, and also to customer service levels.

4.1 Irrigation system performance and operator benefits

Modernization is transforming irrigations systems from labour intensive and data deficient to automatic and data rich. The IMA has established a central 24/7 water operations centre to ensure proper operational management using real time data inputs.

4.1.1 System monitoring and data use

Monitoring of the old manual systems relies on a large field staff presence for visual inspections and manual non-continuous monitoring. Delays in the collection, transmission and analysis of data result in delayed decision making and response to changing circumstances. By contrast, automation under TCC® provides the opportunity to monitor remotely in real time, collect enormous amounts of data and use computer based decision support systems (DSS) that allow water managers to make timely and informed decisions. In addition, automatic alarming allows immediate response to irregular or emergency operational situations. The ability to properly manage extends down to the point of supply to farms. New remote water meters installed as part of the modernization effort now allow the monitoring of service levels at the point of supply to farms and the data gathered allows timely response to customer service issues and problems.

4.1.2 Water savings through better operational control of channels

System generated data (fig. 9) shows just how well demand is matched to capacity in a channel operating under TCC®. The graph is a screenshot of management information data for a typical channel outfall (or escape). Instantaneous and cumulative outfall volumes from the channel (250 ML/d off-take capacity and 18 km in length) are shown for the 2008/09 irrigation season. The channel was modernized in June/July 2008 with FlumeGates™ and operated manually until mid November 2008, when full TCC® was initiated. Cumulative outfalls up until TCC® implementation were 286ML. Outfalls for the remainder of the 2008/09 season (to end April 2009) were just 3ML.

4.1.3 Reduction in labour and improved worker health and safety

Automation provides a fundamental change in field level labour requirements for water operations. Labour once used to physically place and remove timber drop bars and open and close doors and valves is now redundant. The role of labour has changed to one of "field monitoring and maintenance". There has also been a significant improvement in occupational health and safety (OH&S) conditions for workers, with a reduction in work related lost time injuries and improvement in work place safety.

4.1.4 Consistent supply levels

Channel supply levels fluctuate as a consequence of traditional manual regulation, resulting in variable flow rates onto farms. This is a constant source of frustration and complaint by farmers. In order to provide a consistent flow rate, customers must manually adjust the flow through their meter to compensate for the varying supply level.

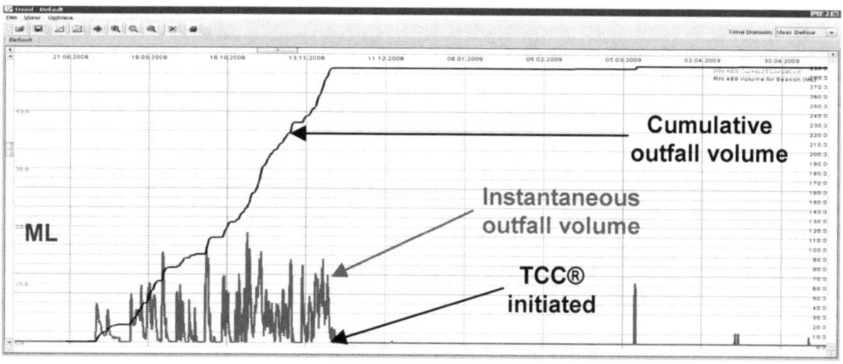

Figure 9: Instantaneous and cumulative outfalls for the CG No 5 channel before and after TCC®.

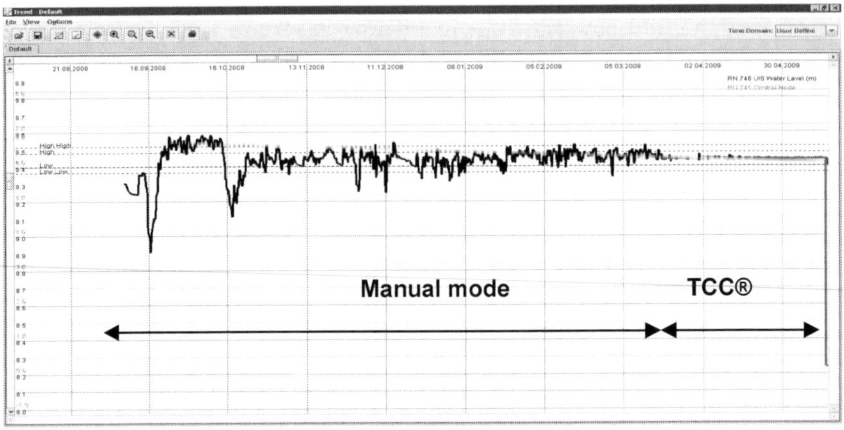

Figure 10: Channel supply level control.

Automation minimises fluctuating supply levels by constantly regulating flows according to demand. FlumeGate™ regulators operate in unison via Rubicon's software program. Figure 10 demonstrates how TCC® provides consistent supply levels compared to operations using traditional manual processes. The FlumeGate™ installed at this site was controlled manually between September 2008 and mid March 2009 when full TCC® was initiated, resulting in a stable supply level in the pool.

4.2 Customer services benefits

Enhanced customer services focuses on improving the flexibility, reliably and equity of irrigation services. The modernized system fundamentally changes the way in which irrigation orders are placed and scheduled and new meters improve the flow rates to farms, resulting in higher volume, shorter duration irrigations. With enhanced operational data and control capabilities, customers are able to make more informed decisions matching the actual crop water requirements on-farm. Orders for water are lodged directly into the irrigation planning database maintained by the IMA [7]. The ordering, scheduling and delivery process is shown in fig. 11. The system is an Interactive Voice Response (IVR) telephone ordering system called WaterLINE, specifically developed by Rubicon Systems for the IMA. Running in parallel with the IVR system is the next generation WEB based water ordering system. This was introduced in 2002, and offers considerable enhancement over the IVR system because of its WEB interface. As of February 2010, over 30% of all orders for irrigation water are lodged via this enhanced WEB based system. It is proving to be a much simpler and clearer management method for IMA staff and customers alike. The internet based system allows customers to view on-screen, the current and future availability of channel capacity at their point of supply. The customer then selects the best available irrigation time to suit his needs.

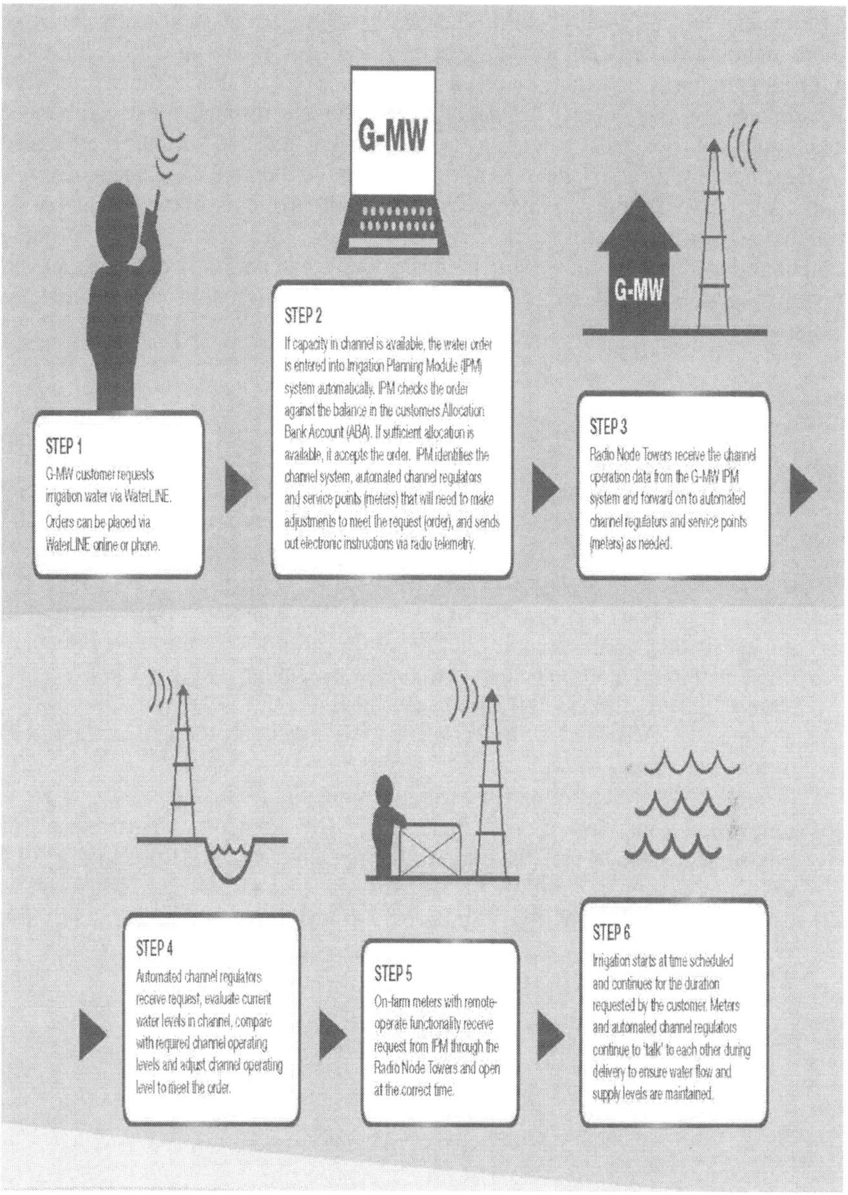

Figure 11: Process steps for placing, scheduling and delivering water. The IMA is Goulburn-Murray Water Authority (G-MW).

5 Conclusions

The experience in Victoria Australia clearly demonstrates that with the right modernization strategy involving a mix of investment in cutting edge technology and infrastructure, large gravity irrigation systems can be made to perform well. Performance improvements are benefiting customers through greatly enhanced customer services and irrigation system managers and through enhanced management information systems and the ability to manage all aspects of water operations in real-time. Water savings brought about by large increases in efficiency of use of water for irrigation means that water saved is being transferred to meet the growing demands in the urban and industrial sectors and to meet environmental requirements and well as improving the security of irrigation allocations.

References

[1] United Nations Population Division www.unpopulation.org

[2] White Paper *Securing Our Water Future Together* , Victorian State Government Department of Sustainability and Environment - June, 2004 ISBN 1 74106 826 6. www.ourwater.vic.gov.au/resources/library/a_z?root_node_selection=4566&result_4567_result_page=S

[3] The Department of Sustainability and Environment (DSE) web-site www.dse.vic.gov.au/dse/index.htm. State water resources planning, implementation and management. (www.dse.vic.gov.au/dse/wcmn202.nsf/Home+Page/592E2077307FBB0CCA256FE100095CDD?open)

[4] Northern Victoria Irrigation Renewals Project (NVIRP) (www.nvirp.com.au)

[5] Rubicon System Australia. www.rubicon.com.au

[6] Innovation in Irrigation, Australian Government, Department of Environment and Water Resources Publication, 2007 ISBN 064255370X www.nht.gov.au/irrigation

[7] Goulburn-Murray Water Authority (G-MWA) Customer Services WaterLINE portal www.g-mwater.com.au/customer-services/waterline

Farmers' preferences for irrigation water demand management tools: a Bayesian analysis of choice behaviour of farmers in Krishna river basin, India

P. Chellattan Veettil, A. Frija, S. Speelman, J. Buysse & G. van Huylenbroeck
Department of Agricultural Economics, Ghent University, Belgium

Abstract

Policy designers have various water demand management (WDM) strategies to achieve efficient water use and allocation at their disposal, such as setting the price of water, decentralizing irrigation water management or improving the quality of water rights. Interestingly, most of these strategies have been applied individually without focusing on the relations between them. This paper uses a discrete choice model to analyse the scope for and the farmer's acceptance of combinations of irrigation WDM tools. In terms of local irrigation water governance the presence or absence of collective irrigation water management among farmers in the form of water user association (WUA) is considered. Water rights are specified by the duration of the title, by their transferability and by the quality of title. Finally, four types of water pricing methods (area, crop, quota and volumetric pricing) are selected for the experimental design. Using the choice experiment we can elicit the most preferred water pricing method under different water rights, water prices and local irrigation water governance contexts. Our results indicated that under conditions of improved water rights, preference for volumetric pricing increases, while the presence of a WUA decreases this preference. We also showed that making the right combination of WDM tools considerably increases the willingness to pay for a change in scenario.

Keywords: complementarity, water pricing method, water rights, local water governance, choice experiment, willingness to pay.

WIT Transactions on Ecology and the Environment, Vol 134, © 2010 WIT Press
www.witpress.com, ISSN 1743-3541 (on-line)
doi:10.2495/SI100031

1 Introduction

Ensuring efficient water use and allocation through the introduction of improved irrigation WDM strategies is a prime concern of many developing as well as developed countries. This should help to reduce the escalating irrigation water demand. To tackle the issue, policy designers adopt different strategies such as setting the optimum price for water rights, decentralizing irrigation water management or improving the quality of water rights. Mostly these strategies are adopted individually or when combined, the relation between them is not checked. Farmers, however, live in a micro-environment under a particular type of irrigation water governance and institutional framework with different cropping systems. The linkages between these items are of the utmost importance. In this paper we, therefore, try to analyze the relations between different WDM tools (water pricing methods and the water rights system) in the context of the Krishna river basin in India. These relations are crucial to improving water resource management, particularly water resource allocation and pricing. For example, water pricing in a poor water rights system is often inefficient and will endanger the cost recovery objective, as well as an efficient water allocation. In situations where water rights are improperly defined, transaction costs increase spectacularly, leading to inappropriate valuation of water resources (Speelman *et al* [1]). According to Liao *et al* [2], the interactions among different water rights and other attributes are very important determinants of irrigation efficiency. In this light, interactions between WDM tools need to be considered while looking at the efficiency of irrigation pricing methods. Moreover, as several authors have shown ([1], Herrera *et al* [3]), willingness to pay for water is affected by the institutional context, which enables us to use contingent valuation in the format of choice experiments to study the complementarities. Because participants in a choice experiment (CE) are required to make a deliberation between different attributes of a single profile, responses are usually less strategic. The choice made by a participant in a CE results from valuation of each attribute solely and their interactions. Therefore, to model the influence of the levels of various attributes and their interactions, a CE is preferred to more conventional contingent valuation methods. To analyse the CE results we use a multinomial probit model (MNP). This model assumes that farmers simultaneously consider all water pricing options while making their choice, which allows us to violate the restrictive substitution pattern (IIA) of the traditionally used multinomial logit model (Train [4]). Using the MNP we can thus assess whether the relative preference of a farmer choosing between two WPMs depends on the presence of other water pricing options.

The remainder of the paper is organized as follows. The second section outlines and briefly discusses the choice experiment setting employed and is followed by a result and discussion section. Finally the paper concludes by highlighting the most important policy implications of the model outcome.

2 Discrete choice analysis

2.1 Study design

In order to analyze the complementary relations between water rights, local irrigation water governance and water pricing, we adopted a choice experiment (CE). One of the initial steps in designing a CE is to identify the choice alternatives and their relevant attributes (Hensher *et al* [5]; Hanley *et al* [6]). Table 1 provides the attributes for the present choice design. These attributes are identified from literature as well as from focus group discussions. Four WPMs are selected as choice options, which are the area pricing method (APM), crop pricing method (CPM), quota pricing method (QPM) and volumetric pricing method (VPM). In APM (status quo), area pricing is employed, which is based on the total acreage of cropping with no restriction of types of crop cultivation. In CPM the water price varies according to the crops grown. In VPM, the unit price of water is fixed and the price of water is calculated by quantity of water consumed times the unit price of water (single rate volumetric pricing), whereas under the QPM, the price of water varies according to the level of water quota allotted to irrigators.

We developed a generic choice design with a calculated D-efficiency of 0.8326 using SAS macros (see Kuhfield [7] for details). The choice design comprise of 16 choice sets, blocked into 4 groups. Each choice set comprise of 4 alternatives with status quo area pricing. Farmers those who do not wish to opt for any of the proposed alternatives can maintain the status quo position. Each block is administered randomly to 80 respondents. An example of a choice set is shown in table 2. The base scenario of the experimental design is area pricing

Table 1: Attributes and levels included in the choice sets.

Attribute	Levels	Remarks
Transfer	1. No transfer	Water rights cannot be transferred even when the entitled water is not used by the farmer
	2. Within WUA	Water rights can be transferred partly or fully, temporarily or permanently to other members within WUA
	3. Between WUA	Water rights are entitled to group (WUA) and transfer of rights would be possible between WUAs
	4. Water market	Water rights can be traded in a water right market
Duration	1. Short (< 2 years)	Water rights entitle to farmers (or WUA) only one or few crop seasons
	2. Long (>2 years)	for longer period so that farmer able to invest on irrigation
Supply	1. Guaranteed 2. Not-Guaranteed	When supply is guaranteed, delivery of water is not interrupted and is according to the contract/title.
Price	4 price levels for each WPM 1.Price I 2.Price II 3.Price III 4.Price IV	Area pricing: 150, 225,300 & 400 Rs/acre Crop pricing: 250, 350, 450, 600 Rs Quota pricing: 360, 600, 1260 & 2100 Rs (quota levels: 30, 50, 70, >70 acre-inch) Volumetric pricing: 8, 12, 18 & 30 Rs.* /acre-inch**

*1 USD ≈ Rs.49.46 at the time of investigation; ** 1 acre-inch = 102.790153 m^3.

with a low price level (150Rs./acre) and short duration water rights with no transaction. An improvement from this scenario will result in financial gain through freedom of choice and acreage of crops grown, higher productivity, reduced total payment for water, higher scope of investments for water saving technologies and reduced risk of crop failure due to water shortage. The financial gain of a farmer can be up to 3-5 times for a farmer cultivating sugarcane in the best scenario (volumetric pricing, long duration water rights, water transferability in market and high price) when compared to a farmer cultivating finger millet at the base scenario. Those who cultivate sugarcane under the base scenario under water stress also expect to benefit in the proposed WPM by reduced water stress, risk and increased productivity.

2.2 Study region and data collection

The Krishna river basin is one of the major river basins of the delta region of India, nourishing three Southern states: Maharashtra, Karnataka and Andhrapradesh. About 77% of the total basin area is cultivable (203,000 Km^2) with an irrigation potential of 47,200 km^2. The majority of the basin area is arid or semi arid and faces high water scarcity.

Primary choice data comprising farmers' choices for WPM were collected from the farmers of Krishna river basin area of northern Karnataka state in India from January to March 2008 by face-to-face interviews. This part includes four sub-basins, namely Lower Krishna, Ghataprabha, Malaprabha and Tungabhadra. The villages and farmers within the villages were selected randomly. Data were collected from 320 farmers. The number of villages and farmers selected is proportional to the total number of villages and farmers in the sub-basins. The details of choice experiment settings were explained to the farmers. Each water pricing method and water attributes were explained in detail. The expected benefit and cost for improvement for attribute level from base scenario to other scenario is explained in terms of choice of crop, increased production, risk and investment opportunities.

2.3 The model

The CE technique relies on the basic idea that an individual can chose (rationally by maximizing utility) among choice sets comprising different attribute levels.

Table 2: An example choice set.

If options crop price, quota price, volumetric price, area price were only available, which one would you choose?

Characteristics	Crop price	Quota price	Volumetric price	Area price (status quo)
Duration	Long	Short	Short	Short
Supply	Not-guaranteed	Guaranteed	Guaranteed	Not-guaranteed
Water transaction	Within WUA	No-Transfer	Between WUA	No-Transfer
Payment	450 Rs	2100 Rs	30Rs/acre-inch	150 Rs
I choose	☐	☐	☐	☐

The utility is derived from the characteristics which a good possess rather than from the goods per se. These *characteristics* or *attributes* are the source of utility for a consumer and they are produced either individually or by combination of goods. Here respondents are asked to choose between different bundles of goods, which are described in terms of their attributes, or characteristics, and the levels that these take.

Assume that utility depends on choices made from some set **C** of alternatives. The representative individual is assumed to have a utility function of the form:

$$U_{ni} = U(Z_{ni}, S_n) \tag{1}$$

where, for any individual *n*, a given level of utility will be associated with any alternative water right characteristic *i*. Alternative *I* will be chosen over some other option *j* if and only if $U_i > U_j$. The utility derived from any option is assumed to depend on the attributes, *Z*, of that option. These attributes may be viewed differently by different individuals, whose socioeconomic characteristics *S* will also affect utility. According to the random utility framework [8], the indirect utility function for each respondent can be partitioned into two parts; one deterministic and in principle observable, and one random and unobservable. Then eqn. (1) can be re-written as:

$$U_{ni} = V_{ni} + \varepsilon_{ni} \tag{2}$$

where $V_{ni} = S'_n\alpha_i + Z'_{ni}\delta$, α and δ are corresponding parameters and the probability that individual *n* will choose option *i* over other options *j* is given by:

$$\begin{aligned} P_{in} &= Pr\left(V_{in} + \varepsilon_{in} > V_{jn} + \varepsilon_{jn}\right) \quad \forall j \neq i \\ &= Pr\left(\varepsilon_{jn} - \varepsilon_{in} < V_{in} - V_{jn}\right) \quad \forall j \neq i \end{aligned} \tag{3}$$

$$= \int_{\varepsilon} I\left(\varepsilon_{jn} - \varepsilon_{in} < V_{in} - V_{jn}\ \forall j \neq i\right)\ f(\varepsilon_n)\ d\varepsilon_n \tag{4}$$

where *I(.)* is an indicator function equalling 1 when the expression in parenthesis is true and 0 otherwise [6]. The probability expression in eqn. (3) is the cumulative distribution of the probability of difference in error ($\varepsilon_{jn} - \varepsilon_{in}$) is below the difference in observed quantity ($V_{in} - V_{jn}$). This give rise to eqn. (4), which is a multidimensional integral over the density of the unobserved portion of utility. The probit model assumes jointly normal error components.

2.3.1 Bayesian estimation of MNP model

A farmer's utility for a WPM *i* is given by eqn. (2) as

$$\begin{aligned} U_{ni} &= S'_n\alpha_i + Z'_{ni}\delta + \varepsilon_{ni} \\ i &= APM, CPM, QPM, VPM \end{aligned}$$

we assume that error terms follow a multivariate normal distribution with mean 0 and variance-covariance matrix Σ, allowing disturbances to be correlated across WPM. In line with random utility models, the farmer chose the WPM which gives the farmer highest utility. Following McCulloch *et al* [8], Imai and van Dyk [9] and Alvarez and Katz [10], we model the observed choice of water pricing method Y_i in terms of utility differentials (latent variable)

$$\widetilde{U}_k = U_k - U_{APM}, \quad k = CPM, QPM, VPM$$

$$Y_n(\widetilde{U}_n) = \begin{cases} 0 & if \max(\widetilde{U}_n) < 0 \\ j & if \max(\widetilde{U}_n) = (\widetilde{U}_{nj}) > 0 \end{cases} \quad for\ n=1,\cdots,N \text{ and } j=1,....,k$$

where $Y_n = 0$ corresponds to the base category APM and N is the sample size.

$$\widetilde{U}_n = W_n\widetilde{\beta} + \widetilde{\varepsilon}_n, \quad \text{with} \quad \widetilde{\varepsilon}_n \sim N_3(0,\widetilde{\Sigma}) \quad \text{and } W_n = \left\lfloor S_n \otimes I_3, Z_n^* \right\rfloor$$

$\widetilde{\Sigma}$ is a positive definite k × k matrix with first diagonal element is fixed ($\widetilde{\sigma}_{11} = 1$) for identification purpose of model parameters $\theta = (\widetilde{\beta},\widetilde{\Sigma})$ in MNP. The likelihood for the multinomial probit model is given by,

$$L(Y|W,\theta) = \prod_{n=1}^{N} \Pr(Y_n|W,\theta)$$

$$P(Y_n|W,\theta) = \int_{Y_n(\widetilde{U}_n)} \phi_3(\widetilde{U}_n|W_n,\widetilde{\beta},\widetilde{\Sigma})\, d\widetilde{U}_n$$

where ϕ_3 is a trivariate normal probability density function.

Bayesian estimation of MNP model was done through Markov chain Monte Carlo simulations using Gibbs sampling algorithm proposed by McCulloch and Rossi (McCulloch *et al* [8], Imai and van Dyk [9], McCulloch and Rossi [11]). This procedure allow us to use the finite sample inferences instead of large sample theory for which a large sample size is required for accurate asymptotic approximations for discrete dependent variables [10,11]. The prior distribution for the model parameters are:

$$\widetilde{\beta}|\widetilde{\Sigma} \sim N(\beta_0, B^{-1}) \text{ and } \widetilde{\Sigma} \sim \text{inv Wishart}(\upsilon,\widetilde{\Delta})$$

where β_0 (=0) and B^{-1} (=0.0001I) are the prior mean and variance of β, υ (= 6) is the degrees of freedom for Σ and Δ (= υI) is the matrix for prior scale of Σ [8]. The joint posterior distribution of the parameters is given by Bayes theorem as

$$k(\widetilde{\beta},\widetilde{\Sigma}|W) \propto L(Y|W,\widetilde{\beta},\widetilde{\Sigma})\,\pi(\widetilde{\beta})\,\pi(\widetilde{\Sigma})$$

The details of the data augmentation algorithm are presented in [9]. The sample obtained after suitable burn in period (where convergence occurred) is regarded as the sample from the target posterior distribution [9]. For estimation, we assume diffuse proper priors for the parameters in the model. Routine sensitivity analyses were done with respect to different priors and starting values, yielding similar results. Convergence of MCMC is diagnosed using Gelman-Rubin [12] statistic. A single Markov chain was run for 1,000,000 cycles, with the first 500,000 were discarded as burn in and a thinning interval of 10 is used. The results presented in section are based on the above 50,000 Gibbs draws of the parameters. The estimation is done using R package *Bayesm*.

2.3.2 Willingness-to-pay (WTP)

WTP estimates are derived by determining the price differences necessary to invoke differences between two attributes or attribute levels (Lusk and Schroeder [13]). Total WTP to obtain water pricing method j versus area pricing is calculated as the ratio of the alternative specific constant to the price coefficient $(-\beta_j/\alpha_j)$. Following [13], marginal WTP for water pricing j versus k can be calculated by subtracting total WTP for water pricing j from total WTP for water pricing k $(-\beta_j/\alpha_j + \beta_k/\alpha_k)$.

3 Empirical results

3.1 Multinomial probit estimates

Table 3 reports the posterior mean and 95% Bayesian credible intervals of the parameters of multinomial probit model. The summaries of posterior densities explicitly point out the relative influence of water attributes and socio demographic variables in the choice of water pricing method. The parameter estimates denote the marginal utility associated with a change from the attribute levels of existing APM. The size of irrigated land significantly influences the choice of quota and volumetric pricing methods. Larger farmers in terms of amount of irrigated land, are more likely to choose quota and volumetric than area pricing method at 95% and 90% level, respectively. However, the size of farm measured as amount of irrigated land does not influence significantly the choice between crop and area pricing method. The dummy variable *price-high* captures the adverse effect of high level of price on choice probability of water pricing method (Alfnes *et al* [14]). A high level of price positively influences the choice of CPM and VPM compared to APM whereas this high level of price negatively and significantly influence the choice between QPM and APM. The insignificance of the coefficient of the *price-high* dummy of VPM explains that the choice of farmers between APM and VPM is not adversely influenced by the high level of price. That is, the stated preference for VPM in the model explains the actual preference of farmers. According to many researchers irrigation water demand is inelastic below a threshold price level, and elastic beyond it (Varela-Ortega *et al* [15]). Volumetric pricing can capture this price elasticity at higher irrigation price level which is explicitly exhibited by the high choice probability for such WPM in our study. Farmers in the presence of WUA are more likely to opt for APM than CPM or QPM or VPM. Interestingly, all the socio-economic variables are significantly affecting the choice of water pricing method. Among the choice specific variables, the duration of water rights and guarantee of supply are not significantly influencing the choice of WPM. All the three levels of water transaction are significantly and positively influencing the choice of WPM compared to No water transaction.

The parameter estimate of price showed that the marginal utility associated with an increase of price by one unit cause a decrease of utility by 1.73 unit. The effect of price on probability of choosing WPM is negative and significant which

Table 3: Parameter estimation of MNP choice model.

Coefficients	mean	Std error	[95% Bayesian credible interval]
Intercept			
Crop/Area	-1.364**	0.828	[-3.3666, -0.128]
Quota/Area	2.142**	0.969	[0.6418, 4.486]
Volumetric/Area	0.277	0.775	[-1.4927, 1.695]
Transaction			
Within WUA	1.214*	0.83	[-0.0176 , 3.193]
Between WUA	2.315**	1.406	[0.2917, 5.749]
Water market	1.325**	0.834	[0.0458, 3.289]
Duration	0.292	0.352	[-0.252 , 1.140]
Supply	-0.306	0.372	[-1.040, 0.465]
Price	-1.735**	0.813	[-3.741 , -0.445]
WUA (dummy)			
Crop/Area	-0.541*	0.29	[-1.086, 0.053]
Quota/Area	-2.881**	1.393	[-6.289 , -0.758]
Volumetric/Area	-1.752**	0.982	[-4.161, -0.339]
Price-high (dummy)			
Crop/Area	0.545*	0.291	[-0.013, 1.125]
Quota/Area	-2.164**	1.28	[-5.245, -0.335]
Volumetric/Area	0.563	0.707	[-0.582, 2.213]
Irrigated land			
Crop/Area	0.018	0.022	[-0.026, 0.063]
Quota/Area	0.071**	0.051	[0.0069, 0.198]
Volumetric/Area	0.055*	0.046	[-0.0051, 0.168]
Variance - Covariance			
Crop/Area: Crop/Area	1	0	
Crop/Area: Quota/Area	0.4486	1.2353	[-2.4476, 2.7957]
Crop/Area: Volumetric/Area	1.1596	1.1196	[-1.1887 , 3.4032]
Quota/Area: Quota/Area	6.1917**	6.9917	[0.3112, 24.9739]
Quota/Area: Volumetric/Area	5.4101**	6.0933	[0.2004, 21.2792]
Volumetric/Area: Volumetric/Area	6.2221**	7.0012	[0.2194, 23.7224]

* and **, significant at 90% and 95% level of Bayesian credible interval.

is in conformity with standard economic theories. The estimates of covariance structure is difficult to explain directly as it replicates the covariance structure of error differences. Hence we rely on substitution pattern of WPM in two conditions: the current situation and an increase of price of area by 25%. If there is no covariance structure exists, the pattern of substitution follow IIA, but Table 4 clearly shows violation of IIA.

There exists a covariance structure in the choice preferences of WPM which makes a difference in percentage shift for different alternatives [6]. When the price of water goes up by 25%, the number of farmers opting for VPM would

Table 4: Substitution pattern of MNP choice model with 25% increase in price.

WPM	Present condition	25% increase in price	Difference in probability (%)	% Change (from current scenario)
Area	0.1938	0.1723	-2.15	-11.09
Crop	0.2406	0.2249	-1.57	-6.53
Quota	0.2085	0.1965	-1.2	-5.76
Volumetric	0.3570	0.4062	+4.92	+13.78

Table 5: Marginal WTP estimates for complementary relations.

Water pricing [a]	Local water Governance [a]	Water rights [a]	WTP (Rs./ acre-inch)
Volumetric pricing	Absent	Poorly defined	3.19
Volumetric pricing	Present	Poorly defined	-17.00
Volumetric pricing	Absent	Well defined	18.47
Volumetric pricing	Present	Well defined	-1.73

([a] Base level: *Area* pricing with the *absence* of local water governance and *poorly* defined water rights)

increase by 14% whereas the all other WPM would decrease. The shifts from CPM and QPM to VPM (6.5 and 5.7% respectively) are nearly equal, whereas it is 11% for APM to VPM (Table 4). This shows an increased acceptance of VPM at higher price levels. It is clear from the substitution pattern that farmers prefer a shift to VPM rather than to CPM or QPM or APM.

3.2 WTP for interactions between WDM tools in VPM

The WTP for interactions between WDM tools are calculated and reported in Table 5. These WTP estimates are in comparison to the base scenario with area pricing, absence of WUA and poorly defined water rights. In all types of relations between pricing, local governance and water rights, the presence of WUA causes a decline in WTP (-17 and -1.73 Rs./acre-inch in case of poorly defined and well defined water rights respectively) when compared to that of similar scenario without the presence of WUA. Thus we conclude that there is a conflicting relation between water governance and pricing as well as between water governance and water rights. The positive relation is evident between water pricing and water rights where as the conflicting relation of presence of WUA reduce the WTP. Ahmad [16] argues that in the absence of well defined water right systems, pricing of water may lead to higher water use instead of reducing it. Additionally, the study of alternative ground water pricing policies in water scarce regions of India by Shiferaw *et al* [17] stressed the importance of local institutions in implementing alternative water pricing policies. They reported a higher acceptance of volumetric pricing based on local experiences such as irrigation charges for a predictable flow per hour. Tsur [18] has mentioned that volumetric based irrigation charges were mainly limited to developed countries due to poor infrastructure facilities available in developing countries. High complementary relations are evident between volumetric pricing and water rights (Rs. 18/acre-inch) when there is no WUA. Thus improvement of water governance structure is necessary to effectively taper complementary relations between water pricing and water rights. World Bank [19] has also reported the difficulties of collecting water payment and has suggested to introduce water rights as well as to set up of local water governance institutions. However, our study demonstrates that presence of such local water institutions can obstruct the efficiency of improved WPM such as VPM.

4 Final remarks

Improved, efficient water pricing is a prime requisite of the rapid growing irrigation sector in India. The introduction of pricing or tax policy for a scarce water resource demands a multidimensional search taking account of different water and allied sectors, stakeholders, socio-demographic, political and geophysical conditions etc. The ideal situation of efficient use of a resource is that proper quantification and administration of the price which is directly applied per unit of undesirable output or limited resource. Therefore, irrigation water pricing involving any kind of volumetric measurement should be the preferred choice. However, transaction cost should not be ignored because they can play an important role in the organization and the control of a pricing system. The present study uses the farmers' perspective to balance between high environmental and economic efficiency and low transactions costs. The result of the study can give ex ante advice about the optimal water pricing and its complementary association with water rights and governance. It is important for farmers to pay only for the water they have actually used especially at high water tariff structure. This can be achieved by volumetric water pricing. The transaction cost depends on water governance as well as water rights and other water institutions. In the current water scarce situation, volumetric pricing is expected to give the best incentive for farmers to increase water use efficiency (Molle and Berkoff [20]).

The cost recovery, efficiency and WDM can be achieved with lower unit price by volumetric based pricing methods. When such pricing method with carefully designed institutional structure is introduced, farmers adopt more efficient and water saving techniques [20]. The WTP estimation showed that volumetric pricing with well defined water rights and improved management of local water governance provide the best result from farmers' perspective. Hence, the focus of the policy should be for VPM with improved water rights and water management. The option of VPM depends on the geophysical and socioeconomic situations in the region and other administrative issues such as decentralisation of water governance, water rights allocation, equity, water availability/supply etc.

References

[1] Speelman, S., D'Haese, M., Frija, A., Farolfi, S. & D'Haese, L. Willingness to pay for water and water rights definition: Study among smallholder

irrigators in Limpopo province, South-Africa. *Proc. of Vth Int. Conf. On Sustainable Water Resources Management*, C.A. Brebbia & V. Popov WIT, Southampton, U.K. pp. 341-352, 2009.

[2] Liao, Y., Giordano, M. & Fraiture, C. An empirical analysis of the impacts of irrigation pricing reforms in China. *Water Policy, 9, pp.* 45-60, 2007.

[3] Herrera, P., Huylenbroeck, G. & Espinel. R. An introduction of the contingent valuation method to assess the efficiency of the institutional structure of irrigation property rights: The case of the peninsula of Santa Elena. *Int. J. of Water Resource Development*, **20 (4)**, pp. 537-551, 2004.

[4] Train, K. *Discrete Choice Methods with Simulation*, Cambridge University Press, Cambridge, 2003.

[5] Hensher, D. A., Rose, J. M. & Greene, W. H. *Applied Choice Analysis: A Primer*. Cambridge University Press, Cambridge, 2005.

[6] Hanley, N., Mourato, S. & Wright, R. E. Choice Modelling Approaches: A Superior Alternative for Environmental Valuation?. *J. of Economic Surveys*, **15(3)**, pp. 435-462, 2001.

[7] Kuhfeld, W. F. *Marketing Research Methods in SAS* Online http://support.sas.com/techsup/technote/ts722title.pdf.

[8] McCulloch, R. E., N. G. Polson, *et al.* A Bayesian analysis of the multinomial probit model with fully identified parameters. *J. of Econometrics*, 99(1), pp. 173-193, 2000.

[9] Imai, K. & van Dyk D. A. A Bayesian analysis of the multinomial probit model using marginal data augmentation. *J. of Econometrics*, **124(2)**, pp. 311-334, 2005.

[10] Alvarez, R. M. & Katz, G. Structural cleavages, electoral competition and partisan divide: A Bayesian multinomial probit analysis of Chile's 2005 election. *Electoral Studies*, **28(2)**, pp. 177-189, 2009.

[11] McCulloch, R. & Rossi, P.E. An exact likelihood analysis of the multinomial probit model, *J. of Econometrics*, 99, pp. 173–193, 1994.

[12] Gelman, A. & Rubin, D. B. Inference from Iterative Simulations Using Multiple Sequences. *Statistical Science*, **7**, pp. 457-472, 1992.

[13] Lusk, J. L. & Schroeder, T. C. Are Choice Experiments Incentive Compatible? A Test with Quality Differentiated Beef Steaks. *A. J. Agri. Econ.* **86(2)**, pp. 467-482, 2004.

[14] Alfnes, F., Guttormsen, A. G., Steine, G. & Kolstad, K. Consumers' Willingness to Pay for the Color of Salmon: A Choice Experiment with Real Economic Incentives. *American. J. Agri. Econ.*, **88(4)**, 1050-1061, 2006.

[15] Varela-Ortega, C., Sumpi, J.M., Garrido, A., Blanco, M. & Iglesias, E. Water pricing policies, public decision making and farmers' response: implication for water policy. *Agricultural Economics*, **19**, pp. 193–202, 1998.

[16] Ahmad, M. Water pricing and markets in the Near East: Policy issues and options. *Water Policy* **2**, pp. 229–242, 2000.

[17] Shiferaw, B., Reddy, V. R. & Wani, S. P. Watershed externalities, shifting cropping patterns and groundwater depletion in Indian semi-arid villages:

The effect of alternative water pricing policies. *Ecological Economics*, **67(2)**, 327–340, 2008.

[18] Tsur, Y. Economic aspects of irrigation water pricing. *Canadian Water Resource Journal*, **30(1)**, pp. 31-46, 2005.

[19] World Bank. *World Bank Water Resources Sector Strategy: Strategic Directions for World Bank Engagement*. World Bank, Washington, DC., 2003.

[20] Molle, F. & Berkoff, J. Water Pricing in Irrigation: Mapping the Debate in the Light of Experience. *Irrigation water pricing: The gap between theory and practice*, eds. F. Molle & J. Berkoff, CAB International, UK., 2007.

Effect of deficit irrigation and mulch on water use and yield of drip irrigated onions

A. A. Ramalan[1], H. Nega[2] & M. A. Oyebode[1]
[1]*Department of Agricultural Engineering, Ahmadu Bello University, Nigeria*
[2]*Federal Ministry of Agriculture and Rural Development, Ethiopia*

Abstract

A field experiment was conducted at Melkassa Agricultural Research Centre in the Central Rift Valley of Ethiopia (located at latitude 8° 24'N, longitude 39° 21'E and altitude 1,550 m above sea level) during the 2008/2009 dry season to study the 'Effect of deficit irrigation and different mulch covers on water use and yield of drip irrigated onion (Allium Cepa L., cultivar: Bombay Red)'. A drip irrigation system with 144 laterals was used for the experiment. Two factors, namely deficit irrigation at four levels (0, 25, 50 and 75% of Total Available Water, TAW) and mulch covers of four different types (no mulch or bare soil, straw mulch, clear plastic and black plastic) were combined in a randomized complete block design and laid out as treatments in plots 0.4 m x 5 m. Three replications were used. Results showed that onion bulb yield decreased with increase in levels of water deficit. In contrast, both crop water use and irrigation water use efficiencies increased with increase in water deficit level. Use of mulch significantly increased bulb yield. Plots with mulch covers yielded higher than 30 $t.ha^{-1}$compared to the bare soil yield of 28.3 $t.ha^{-1}$ (straw>clear plastic>black plastic>bare soil). Water use was similarly reduced by the presence of mulch covers. A hydraulic performance evaluation of the drip system showed there was an average emitter flow rate variation of 13.7%, a coefficient of uniformity of 85% and an overall irrigation application efficiency of 94.4%.

Keywords: deficit irrigation, drip irrigation, water harvesting, mulch, irrigated onion, hydraulic performance of drip system, water use, water use efficiency.

doi:10.2495/SI100041

1 Introduction

The dry land area of Ethiopia which includes the Central Rift Valley accounts for 67% of the total land mass of the country (EARO [5]). Approximately, 62% of the irrigable land of the country is located in this Valley. The major crop irrigated by farmers in the area, who are mainly smallholder and peasant, is onion. Rainfall, characteristically high intense and capable of causing erosion, rarely exceed 800 mm; most areas receive 500-800 mm annually (MOA) [12]. Annual potential evapotranspiration is high, thus placing high evaporative demand on water surfaces. As a strategy for reducing food insecurity, encouraging small-holder farmer investment in agriculture and increased exploitation of its water resources, the government introduced rain-water harvesting and small scale drip irrigation technologies. In this regard, over 100,000 units of water harvesting structures, hand-dug wells, low cost water lifting devices, and family drip irrigation systems deployed have achieved partial success in implementation. Knowledge gaps exit regarding the correct sizing of water harvesting tank, irrigation land size relative to household tank, effective use of harvested water and the economics of the whole system. These have to be addressed in a study. For a general objective, this pioneering study sought to evaluate the effect of deficit irrigation and use of mulch covers on water use and yield of drip irrigated onion targeting resource poor smallholder farmers in moisture deficient environment.

1.1 Deficit irrigation effects

Deficit (or regulated deficit) irrigation scheduling is one way of maximizing water use efficiency for higher yields per unit of irrigation water applied; the crop is exposed to a certain level of water stress either during a particular period or throughout the whole growing period (FAO [8]). The expectation is that any yield reduction resulting from the water stress will be insignificant compared with the benefits gained through diverting the saved water to irrigate other crops. For deficit irrigation and its schedules to be successful, some knowledge of crop response to water stress at specific growth stage or entire season is essential. FAO [8] suggested that crops, or crop varieties most suitable for deficit irrigation are those with short growing season and are tolerant to drought. In implementing deficit irrigation, consideration must be given to soil retention capacity, as well as, modification of agronomic practices like plant population, date of planting, fertilizer application, etc.

Crop water use efficiency is the ratio of crop yield (Y) to the amount of water depleted by the crop in the process of evapotranspiration (ET) and is expressed as $Kg.m^{-3}$ or $Kg.ha^{-1}mm^{-1}$. Irrigation water use efficiency, on the other hand, is a ratio of the crop yield (Y) to the total amount of water used in the field (WR). Since both the Crop and Irrigation water use efficiencies are ratios, their values can be increased by either increasing crop yield through better method of pest and disease control, plant breeding or decreasing evapotranspiration. Bekele and Tilahun [2] reported the positive effect of water stress at two growth stages of onion. In contrast, Nigatu [16] reported no significant improvement in irrigation water use efficiency resulting from irrigation deficit. Under conditions of water shortage, the best growth stages at which to achieve maximum water savings in the irrigation of onion (cv: Bombay Red) without yield reductions exceeding 31% are development (bulb formation) and maturity stages at which 50% of water requirement is applied.

1.2 Mulch cover effects

Mulches can be composed of plant materials or they may be synthetic mulches consisting of plastic sheets (Allen et al. [1]). The sheets are generally transparent, white or black. Colour as observed by Allen et al. [1] influences albedo mainly during the early stage of crops. Plastic mulches substantially reduce evaporation of water from the soil surface especially under trickle/drip irrigation. Among its advantages in use is increase in soil temperature, reduced fertilizer leaching, reduced evaporation, cleaner product and reduced weed problems. Higher initial cost and management are some of its disadvantages. Studies reported in Allen et al. [1] indicated that the average approximate reductions in crop coefficient, Kc and surface evaporation for five horticultural crops (squash, cucumber, cantaloupe, watermelon and tomato) under complete plastic mulch compared with no mulch using drip irrigation system was 10-30% and 50-80% respectively. Douglas [4] reported yield increases over state average of four times for pepper, squash and watermelon; and three times for tomato when plastic mulch and drip irrigation were used.

1.3 Drip irrigation system hydraulics

Drip (or sometimes referred to as trickle) irrigation system derived its name from the manner in droplets reaches the soil surface. Michael [14] reported it as one of the latest of the systems and is become increasingly popular in areas with water scarcity and salt problems. Water from source passes through plastic pipes, constituting the main and laterals, into emitters positioned to supply each plant with the calculated water requirement at the same delivery rate. Pressure head losses are encountered in lines which result in uneven distribution of the discharges from the emitters. Mofoke et al. [15] reiterated that the most widely accepted hydraulic performance indices for assessing the drip irrigation system are emitter discharge, emitter flow rate variation, uniformity coefficient and emission uniformity.

2 Materials and methods

2.1 Climatic and soils description of the study area

The climate of the study area (Mekassa: located at latitude 8° 24'N, longitude 39° 21'E and altitude 1,550 m above sea level) is characterized as semi arid, having a long term (1977-2006) mean annual rainfall of 791.7 mm. The rainfall has a bimodal pattern with a short rain season from March to April, a dry spell, and a long rain season from July to September. The long term mean maximum and minimum temperatures range from 30.9°C in May to 26.2°C in August and from 16.2°C in June to 10.8°C in December respectively. Recorded mean maximum and minimum temperatures which occurred in October and December were 28.18°C and 8.19°C respectively. The highest duration of sunshine was 10.4 hours in December with the lowest of 6.34 hours in October. The mean relative humidity was highest in October at 57.42% and lowest at 44.71% in December. Wind speed measured with the cup anemometer mounted at 2 m height recorded the highest mean wind speed of 10.04 $m.s^{-1}$ in December while the lowest mean wind speed of 6.34 $m.s^{-1}$ was in October. Pan evaporation was highest in December with a mean value of 7.27 mm.

2.2 Soil characteristics at site

The soil at the site is classified as Haptic Andosol with a characteristic feature of deep pumice or volcanic soil type having a high water retention capacity (Tesfaye and Walker, 2004) as cited by Nigatu [16]. A composite sample was collected at depth of 0-60 cm from the site and analyzed for physical and chemical characteristics at the Ethiopian National Soils Laboratory. The results show that the soil is loam in texture with a mean bulk density of 1.11 $gm.cm^3$ and slightly alkaline with a pH of 7.89 and an EC of 0.208 $dS.m^{-1}$. The soil was found suitable for irrigation. The soil was also assessed for infiltration using the Double Ring Infiltrometer.

2.3 The experiment

The two factors for investigation in the experiment were deficit irrigation at four levels (0, 25, 50 and 75% of Total Available Water, TAW) combined in a randomized complete block design with four types of mulch covers (no mulch, *tef* straw mulch, black and clear plastic mulches). These constituted sixteen treatments each laid in plots 1.8m x 5m. Each treatment was replicated three times, totalling forty eight plots covering a gross area of 0.086 ha.

2.3.1 Field operation

Onion seeds (Allium Cepa, L. Cv: Red Bombay) at the rate of 3 $kg.ha^{-1}$ were sown on six seed beds 1m x 5m to raise the seedlings required at a nursery close to the site on August 28, 2008. Nursery bed management according to the recommendations of the Experiment Station were followed. At seven weeks after sowing when the seedlings were about 10-15 cm high and had three true leaves, they were transplanted to 40 cm wide planting beds at intra-row and inter-row spacings of 10 cm and 20 cm respectively. Prior to transplanting, soil moisture content at depths up to 30 cm were determined using the gravimetric method and one irrigation applied to raise the moisture content of the soil one day before transplanting to field capacity level. Transplanting was done on October 20, 2008. After transplanting, full irrigation to restore soil moisture content to field capacity, based on effective rooting depth of 15 cm, was given to all treatment plots. On 11 November, 2008, the irrigation and mulch treatments were imposed.

Fertilizers and pest control chemicals were applied based on recommendations of the Research Station. Urea at the rate of 100 $kg.ha^{-1}$ was applied between rows in two equal doses at the time of transplanting and at 4 WAT. A uniform dose of Diammonium phosphate (DAP) at the rate of 200 $kg.ha^{-1}$ was applied and incorporated in the soil before transplanting. Pesticides chemicals were applied at 3, 4, 5 and 8 WAT. Three hand-hoe weeding were carried out at 4, 6 and 10 WAT. The crop was harvested 18 WAT on January 31, 2009 when the tops of more than 50% of plants had fallen over.

2.3.2 Irrigation water source

Rainwater harvested from farm roads and stored in a 50 m^3 capacity underground tank, 6m deep, was the main source of the irrigation water. A 50 mm diameter petrol engine pump was used to lift water from the underground tank to a 1.5 m^3 capacity elevated tank, 2.5 m above ground. Pump capacity was calculated based on tank capacity and filling time.

2.3.3 Drip system components

Water from the elevated tank feeds into a supply line 20 mm diameter, 5 m long made from Low Density Polyethylene Pipe (LDPEP). A ball valve and a primary filter are fixed on the line and it terminated at a 20 mm, 19 m long mainline of the same material. Four sub mainlines each 34.6 m long, 20 mm diameter were connected to the mainline. Along each of the sub mainline, six tertiary lines each measuring 4.6 m, 20 mm were connected and it was on these tertiary lines that 12 mm diameter, 5m long, 70 cm spaced laterals with inline emitters took off. There were 144 laterals altogether installed. The hydraulic characteristics of the system installed that were evaluated included: emitter flow rate, emitter flow rate variation, uniformity coefficient, emission uniformity, and application efficiency.

2.3.4 Hydraulic characteristics of the system installed

2.3.4.1 Emitter flow rate The average flow rate of the emitters used in the experiment was measured from plots in which catch cans were randomly assigned plots and volumes of flow caught over a time period. The discharge, or flow rate out of single outlet emitter at a specified head was estimated thus:

$$q = \frac{V}{\Delta t} \tag{1}$$

where q = single emitter discharge, litre/hour; V = volume of water collected from emitter, litres; and Δt = time duration of discharge collection, hour

2.3.4.2 Emitter flow rate variation Emitter flow rate variation was simply computed as:

$$q_{var}(\%) = 100\left(\frac{q_{max} - q_{min}}{q_{max}}\right) \tag{2}$$

where q_{var} = emitter flow rate variation, %; q_{max}= the maximum emitter flow rate along a lateral, litre/hour; and q_{min} = the minimum emitter flow rate along a lateral, litre/hour.

2.3.4.3 Uniformity coefficient Uniformity coefficient was calculated using the Christiansen coefficient of uniformity formula given in Michael [14] as:

$$UC(\%) = 100\left(1.0 - \frac{\Sigma_x}{qn}\right) \tag{3}$$

where UC(%) = Uniformity coefficient, %; q = average discharge of the emitters, litre/hour; n = number of emitters; and $\sum x$ = sum of the individual deviations of observed flow from the average discharge, litre/hour.

2.3.4.4 Emission uniformity The formula given by Howell and Hiller [9] cited in Michael [14] as used to estimate emission uniformity as:

$$Eu(\%) = 100\left(\frac{q_{\text{low quarter}}}{q}\right) \tag{4}$$

where Eu = Emission uniformity in %; $q_{\text{low quarter}}$ = average discharge rate of the low quarter of the number of emitters observed, litre/hour; and q = average discharge rate of all observed emitters, litre/hour.

2.3.4.5 Application efficiency The overall application efficiency of the drip irrigation system was estimated from the relationship of Vermeiren and Jobling [18] given as:

$$Ea(\%) = Ks.Eu \tag{5}$$

where Ea = Application efficiency, %; Ks = Average water stored in the root zone over average depth of water applied and is a coefficient which expresses the storage efficiency of the soil taking into account the pressure variation in the drip system (Ks=1 for loam soil) as in Vermeiren and Jobling [18] and Eu = as given in eq. (4).

2.3.5 Water use and water use efficiencies

Soil water use in the experiment was obtained principally from routine measurements of soil moisture content by the gravimetric method and use of Neutron probe meter. For the soil depth of 0-15 cm, soil samples were collected with soil auger just before irrigation and one day after irrigation to compute soil water contents as:

$$SW = \Theta v.Z = \Theta m.Bd.Z = \frac{(FW - DW)}{DW}.Bd.Dz \tag{6}$$

where SW = Soil water content in the soil layer (0-15 cm), cm; Θv, Θm = volumetric and gravimetric moisture contents in the depth in cm^3/cm^3 and g/g respectively; Dz = depth of soil layer, cm; Bd = Bulk density of the soil, g/cm^3; and FW, DW= Wet and dried weights of soil sample, g.

For soil layer below 15 cm depth, the Neutron Probe meter (Model: 503 DR CPN Hydro Probe) with forty eight access tubes made of 5 cm diameter, 150 cm long aluminium pipes were used to obtain soil water content up to depths of 40 cm in increments of 7.5 cm. A calibration equation using the method in Eze et al. [6] was programmed into the probe so as to obtain direct readout moisture content in mm.

Total available water is the water held in the root zone depth at moisture contents defined as field capacity (upper limit) and permanent wilting point (lower limit). In the experiment, upper limits of soil moisture contents were fixed at 0, 25, 50 and 75% of TAW. Water use by the onion was therefore calculated based on change in soil moisture. The design of drip system predicated on zero runoff and deep percolation losses. The net depth of irrigation water applied was based on the moisture depleted from storage to satisfy actual crop evapotranspiration minus any rainfall in the interval plus leaching fraction (leaching was not considered in the study as the soil was non-saline). Effective rainfall was accounted for by the method suggested by FAO [7]. Gross depth of irrigation was calculated taking into account application efficiency of the system. Water use efficiency is a measure of crop water productivity which Michael [14] suggested can be measured in the form of:

$$CWUE = \frac{Y}{CWU} \tag{7}$$

where CWUE = Crop water use efficiency, $kg.m^{-3}$; Y = Crop yield, $kg.ha^{-1}$; and CWU = Crop water use, mm

and

$$IWUE = \frac{Y}{IWU} \tag{8}$$

where IWUE = Irrigation water use efficiency, $kg.m^{-3}$; Y = Crop yield, $kg.ha^{-1}$; and IWU = Irrigation water use, $m^3.ha^{-1}$.

3 Results and discussion

3.1 Emitter flow rate and flow rate variation

The results in Table 1 show emitter flow rate as affected by the number tested at a time using the catch can test method. At an operating pressure head of 2.5 m, the average emitter flow rate was 0.6045 litre/hr while mean maximum and mean minimum values were 0.656 and 0.566 litre/hr respectively. The average obtained, though at a higher pressure head, was lower when compared with the manufacturer's specification of 1.3 litre/hr at a pressure head of 1.0 m. The mean coefficient of variation obtained for the results was 4.60, indicating relatively low variations of the mean.

Table 1: Average, maximum and minimum emitter flow rates as affected by the number of treatments tested at a time using catch cans.

Parameters	Number of treatment plots tested at a time					
	48	24	18	9	3	Average
Average emitter flow rate, litre/hr	443.4	624.9	663.6	648.2	642.2	604.5
Maximum emitter Flow rate, litre/hr	480.0	670.0	760.0	680.0	680.0	656.0
Minimum emitter flow rate, litre/hr	410.0	580.0	620.0	610.0	610.0	566.0
Coefficient of variation, C.V. %	5.10	5.40	5.80	3.80	3.80	4.60

3.2 Emitter flow rate variation, uniformity coefficient, emission uniformity and application efficiency

The average emitter flow rate variation along laterals was 13.7%, as shown in Table 2. This value closely agreed with the findings in Michael [14] and Jensen [11] which stated that in drip systems, the average variation in discharge rate of individual emitters in a whole field should not exceed 20%.

Table 2: Emitter flow rate variation, uniformity coefficient, emission uniformity and application efficiency as affected by the number of treatments tested at a time using catch cans.

Parameters	Number of treatment plots tested at a time[a]					
	48	24	18	9	3	Average[b]
Emitter flow rate variation, %	14.6	13.4	18.4	11.6	10.3	13.7
Uniformity coefficient, %	95.6	95.7	95.3	96.4	96.8	96.0
Emission Uniformity, %	93.1	93.6	94.0	95.0	96.0	94.4
Application Efficiency, %	93.1	93.6	94.0	95.0	96.0	94.4

[a]While testing specific treatments, gate valves for the others are closed.
[b]Each treatment had 3 plots due to replication.

Over the five separate tests, the average emission uniformity was found to 94.4% while the mean of the uniformity coefficient was 96%, signifying even distribution of water throughout the system. These results are supported by Merriam and Keller [13] who stated that a drip system with both emission uniformity and uniformity coefficient of 85% or greater, and with a discharge variation of less than 20% should be considered satisfactory.

The application efficiency obtained averaged 94.4%; this compared favourably to the results of Sijali [17] and IDN [10] who stated that drip irrigation systems are typically about 90% efficient as compared to sprinklers that are about 75% efficient.

3.3 Deficit irrigation and mulch effects on bulb yield and seasonal water use

The highest marketable bulb yield of 37.2 $t.ha^{-1}$ was obtained at the 0% deficit irrigation level and the lowest mean yield of 24.32 $t.ha^{-1}$ recorded at 75% level of irrigation deficit as presented in Table 3. The yield differences were significant at 1% level of probability. The yield reductions with increased deficit irrigation levels from 0 to 25, 50 and 75% were 16, 26, and 34.6% respectively which resulted from corresponding reductions of applied irrigation water of 14, 29 and 47%.

Effect of deficit levels on Crop water use and irrigation water use revealed significant differences in seasonal values of the water use. In a nutshell, allowing for no deficit, an onion crop in the Central Rift Valley requires 527.18 mm of irrigation water per season to yield 37.2 $t.ha^{-1}$. But, imposing a deficit level of 75% TAW the yield obtainable drops to 24.32 $t.ha^{-1}$. These results, up to deficit levels of 50%, tally with Doorenbos and Kassam [3] report which stated that for an optimum yield, onion requires 350-550 mm of water.

Use of mulch had resulted in significant yield differences (p=5%). Onion yield under straw mulch, though at par with the plastic mulches, out-yielded bare soil (no mulch). No yield advantage was observed using clear as against black mulch. Seasonal water uses were significantly reduced under mulches (straw>clear plastic>black plastic) than with bare soil which resulted in over 400 mm per season.

Though the data showed no significant differences resulting from the interaction of irrigation deficit and mulch covers, the combination I_oM_1 (0% deficit level, straw mulch) yielded highest, close to 40 $t.ha^{-1}$.

3.4 Deficit irrigation and mulch effects on water use efficiencies

Crop and Irrigation water use efficiencies were highest at the 75% irrigation water deficit at which 14.52 and 13.50 kg/m^3 were obtained respectively as Means followed by the same letter(s) in a column of any treatment group are not statistically different at the 5% (*), or 1% (**) probability level using the Duncan Multiple Range Test. ns=not significant.

I_o=0% deficit, I_1=25% deficit, I_2=50% deficit, I_3=75% deficit, M_o=no mulch, M_1=Straw mulch, M_2=clear plastic mulch, M_3=black plastic mulch.

CWUE=Crop water use efficiency, IWUE=Irrigation water use efficiency.

Table 3: Onion bulb yield, crop water use, irrigation water use, CWUE and IWUE as affected by deficit irrigation and mulch covers at Melkassa in the 2008/2009 dry season.

Treatments	Marketable bulb yield ($t.ha^{-1}$)	Crop water use (mm /season)	Irrigation water use (mm/season)	CWUE (kg/m^3)	IWUE (kg/m^3)
Deficit irrigation levels (%)					
I_o	37.20a	490.28a	527.18a	11.51b	10.70b
I_1	31.23b	420.38b	452.02b	11.20b	10.41b
I_2	27.39c	349.25c	375.53c	11.86b	11.03b
I_3	24.32d	259.73d	279.28d	14.52a	13.50a
Significance level	**	**	**	**	**
Mulch covers					
M_o	28.27b	417.77a	449.21a	10.29c	9.57c
M_1	31.71a	353.25b	379.84b	13.89a	12.92a
M_2	30.12ab	364.84b	392.30b	12.78ab	11.88ab
M_3	30.04ab	383.77b	412.66b	12.13b	11.28b
Significance Level	*	*	*	**	**
Interaction (deficit x mulch)					
I_oM_o	34.94	535.84	576.16	9.85d	9.16d
I_oM_1	39.46	449.60	483.44	13.17bc	12.25bc
I_oM_2	37.02	462.69	497.52	12.08cd	11.13cd
I_oM_3	37.38	512.97	551.58	10.93cd	10.17cd
I_1M_o	30.24	454.39	488.59	10.00d	9.30d
I_1M_1	32.86	402.40	432.69	12.28cd	11.42cd
I_1M_2	31.19	406.79	437.41	11.48cd	10.67cd
I_1M_3	30.65	417.92	449.38	11.03cd	10.26d
I_2M_o	26.84	365.40	392.90	11.10cd	10.33cd
I_2M_1	28.63	334.01	359.15	12.97bc	12.06bc
I_2M_2	27.11	343.70	369.57	11.87cd	11.04cd
I_2M_3	26.96	353.87	380.50	11.49cd	10.69cd
I_3M_o	21.07	315.47	339.21	10.20d	9.49d
I_3M_1	25.89	226.97	244.06	17.14a	15.94a
I_3M_2	25.14	246.17	264.70	15.68a	14.58a
I_3M_3	25.18	250.32	269.16	15.06ab	14.00ab
Significance Level	ns	Ns	ns	*	*
CV (%)	7.20	9.68	9.68	11.1	11.1

CWUE and IWUE. The other levels were statistically equal. Singular effect of use of mulch showed the straw mulch as the best material, though at par with clear plastic (1% level). Bare soil (no mulch) was not efficient in conserving water. The interaction of deficit irrigation and mulch showed there were significant differences in efficiency values when tested at the 5% level. At I_3M_1, the highest CWUE of 17.14 kg/m^3 and IWUE of 15.94 kg/m^3 were obtained. The least was at I_0M_0. These results show the positive effect of mulch covers even with deficit levels up to 75%.

4 Conclusions

Results from the present level of work show that combining irrigation deficit with use of mulch covers, particularly straw mulch that is in abundance with farmers, presents a sustainable strategy for onion production in the semi-arid areas of Ethiopia. Use of plastic mulches did not show any superiority over straw mulches; laying difficulties with these materials in windy conditions present challenges to farmers. Imposing irrigation deficit levels beyond 50% is not advisable as marketable bulb yield and bulb sizes decline. The drip system technology used in the study is at entry level for farmers and it is aptly demonstrated from this work that the amount of water savings that can be achieved is substantial when compared to other irrigation systems.

Acknowledgements

The authors of this paper wish to acknowledge with thanks the support provided by Sasakawa Africa Fund for Extension Education (SAFE). Similar thanks go to the Agricultural Extension Department of the Federal Ministry of Agriculture and Rural Development of Ethiopia for approving the study leave of the second author. The assistance rendered by Melkassa Agricultural Research Centre and the National Soil Testing Centre in the form of use of research land, field staff, equipment and analysis at no cost is very much appreciated.

References

[1] Allen, R.G., Pereira, L.S., Raes, D. & Smith, M. *Crop evapotranspiration.* Guidelines for computing crop water requirements. FAO Irrig. Drain, Paper No. 56. FAO. Rome, Italy. 300pp. 1998.

[2] Bekele, S. & Tilahun, K. *Regulated deficit irrigation scheduling of onion in a semi-arid region of Ethiopia.* Agricultural water Management. **89:** pp.148-152, 2007.

[3] Doorenbos, J. & Kassam, A. *Yield response to water.* Irrigation and Drainage Paper No. 33. FAO. Rome, Italy. 193pp. 1979.

[4] Douglas, C.S. *Using plastic mulches and drip irrigation for vegetable production.* http://www.ces.ncsu.edu/depts/hort visited 06/12/2007.

[5] EARO. *Dry land Agricultural Research Strategic Plan*, Addis Ababa, Ethiopia. 1999.

[6] Eze, P.C., Alhassan, A.B., Kundiri, A.M. & Mamman, E. Neutron moisture calibration – a tool for effective soil moisture monitoring. *Proc. of the 13th National Irrigation and Drainage Seminar.,* eds. I.U. Abubakar & M.K. Othman, Ahmadu Bello University Press: Zaria Nigeria, pp. 232-235, 2007.

[7] FAO. *Irrigation water management:* Irrigation water needs. Training Manual No. 3. Rome, Italy. 1986.

[8] FAO. *Deficit Irrigation Practice,* Water Report 22. Rome, Italy. 88p. 2000.

[9] Howell, T.A. & Hiller, E.A. Optimization of water use efficiency under high *frequency irrigation* – I. Evapotranspiration and yield relationship. Transaction: ASAE, **15(5)**, pp. 873-878, 1975.

[10] IDN (International Development Enterprise). Technical Manual for affordable micro irrigation technology. http://www.ide.international.org visited 14/06/2007.

[11] Jensen M.E. (ed.) *Design and operation of farm irrigation systems*, ASAE Monograph 3, American Society of Agricultural Engineers, St Joseph, Michigan, pp. 447–500. 1983.

[12] MOA. *The five year action plan of soil and water conservation activities.* Forest and Wildlife, Soil and Land Use Technologies Regulatory Department, Federal Ministry of Agriculture, Addis Ababa, Ethiopia. 2001.

[13] Merriam, M. & Keller, J. *Farm irrigation system evaluation: a guide for Management.* Printed in the United States of America, Logan, Utah. 1978.

[14] Michael, A.M. *Irrigation: Theory and Practice.* Vikas Publishing House Pvt. Ltd. New Delhi, 801pp. 1978.

[15] Mofoke, A.L.E., Adewumi, J.K., Mudiare, O.J. & Ramalan, A.A. *Design, contruction and evaluation of an affordable continuous-flow drip irrigation system.* http://www.vl-irrigation.org visited 6/11/2007.

[16] Nigatu, A.W. *Identification of the critical water requiring growth stage on Onion (Allium cepa L.) for bulb production in the Central Rift Valley.* MSc. Thesis. Haramaya University. Ethiopia. 2008.

[17] Sijali, I.V. *Drip irrigation: Options for smallholder farmers in eastern and southern Africa.* RELMA Technical Handbook Series 24. Nairobi, Kenya. 2001.

[18] Vermeiren, L. & Jobling, G.A. *Localized irrigation: design, installation, operation and evaluation.* Irrigation and Drainage Paper 36, FAO, Rome, Italy. 1980.

Magnitude yield response and economic value of selected wheat genotypes related to irrigation schedules under the arid ecosystem of Saudi Arabia

A. A. Alderfasi[1] & A. A. AL-Owayed[2]
*[1]Plant Production Department,
Faculty of Food and Agricultural Sciences, King Saud University,
Saudi Arabia*
*[2]Biology Department, Faculty of Sciences,
The Princess Noura Bint Abdelrahman University for Girls, Saudi Arabia*

Abstract

The greatest challenges facing the improvement of agriculture programs under arid and semiarid environment are those involving the provision of water. Thus, field experiments have been carried out in a split–plot design at the Agricultural Research Station, Faculty of Food and Agricultural Sciences, King Saud University. The experiments included 20 treatments, four wheat genotypes, (Yecora Rojo, West bread, KSU 102 and KSU 105) and five irrigation schedules (50, 100, 150, 200 mm) of cumulative pan evaporation CPE, as compared to traditional irrigation methods used by many farmers (weekly irrigation). The total amount of water supplied over the growing season was calculated. The data obtained clearly indicated that a gradual decrease in most of the yield and yield component characters were in line with decreasing irrigation schedules. The results also showed that no significant differences were found between irrigation schedules at 50 mm of CPE and weekly irrigation in most of the studied characters, except on biological yield, which was similar to traditional irrigation. In addition, significant differences among wheat genotypes were observed in all traits. The KSU 105 genotype ranked in the first position and surpassed the other tested genotypes. The interaction between irrigation schedules and genotypes was also significant for grain yield. Economic evaluation showed that the highest invested return per SR and net return in SR was obtained by irrigating wheat at

WIT Transactions on Ecology and the Environment, Vol 134, © 2010 WIT Press
www.witpress.com, ISSN 1743-3541 (on-line)
doi:10.2495/SI100051

50 mm, followed by 100 mm CPE. Finally, we can conclude that at least 20% of water irrigation could be conserved to achieve high grain yield.

Keywords: wheat genotypes, arid environment, drought resistance, irrigation schedules, CPE, economic value.

1 Introduction

Interest in crop response to environment stress has increased greatly in recent years because severe losses may result from heat, cold, drought and high concentrations of salts in the available water used in irrigation. On a global basis, drought (assumed to be soil and/or atmospheric water deficit), in conjunction with coincident high temperature and radiation, poses the most important environmental constraint to plant survival and crop productivity. Several studies indicated that wheat grains showed a high response to soil moisture stress. The magnitude of yield reduction from water deficits in wheat genotypes depends upon the growth stage at which the water deficiency occurs and the severity and duration of the deficiency (Salter and Goode [1]; Misra et al [2]; Chauhan et al [3] and Abd el Ghany [4]).

Saudi Arabia, like many other countries located in the arid and semiarid regions of the world, faces the problem of limited water resources (Hussain and Al-Jaloud [5] and Ghandorah et al [6]). Wheat is a major crop of such areas. Although wheat is considered as being fairly tolerant to drought, its yield is drastically reduced due to low soil moisture (Kramer and Boyer [7] and Alderfasi et al [8]). Moisture stress at any stage of crop growth can cause an irreversible loss in yield potential. The severity of loss depends on many factors, such as timing, length and severity of the drought period. Moreover, yield reduction can be due to the reversible effects in the number of tillers, reduced kernel weight or fewer kernels (Reginato [9] and Reginato and Carrot [10]). Under such conditions it was necessary to develop efficient reliable and economical irrigation management strategies for effective use of the existing limited water resources (Imtiyaz et al [11, 12]). The traditional method for irrigation scheduling that depends on time intervals is not efficient, and improper irrigation management practices do waste scarce water resources. Therefore, it is essential to develop these methods through untraditional ways. One of these efficient reliable methods is cumulative pan evaporation (CPE). Using the ratio between irrigation water applied and CPE has been accepted by many researchers due to its simplicity and data availability [12–14]. Due to lack of proper irrigation scheduling techniques, the average yield of wheat and other field crops is low either due to excess or deficit soil moisture regimes. In respect of genotypic variation effect, several investigators have found that this appears to give great response, since specific ones surpassed others significantly in wheat yield and its components (El-Haddad et al [15], Yousef and Hanna [16], El-Beially [17], Hassan et al [18], Abd Allah et al [19] and Reiad et al [20]). The importance of water stress genotypes could be more accurately predicated with a clear picture of the relationships between growth stages and plant response to water stress. Keeping in mind the above mentioned review of literature, the present study was proposed to clarify the relationship between yield productivity and water schedules for some wheat genotypes. In addition, economical evaluation was planned as an additional target.

2 Materials and methods

2.1 Plant materials

Four wheat genotypes (two cultivar varieties Yecora Rojo and West bread and two advanced lines KSU 102 and KSU 105) were tested in a field experiment. The pedigree and identification of the genotypes used are presented in Table 1.

2.2 Water irrigation schedules

Traditional irrigation used by many farmers (weekly irrigation) plus four irrigation schedules at cumulative pan evaporation (CPE) of 50, 100, 150 and 200 mm during the entire crop growth period with the depth of 50 mm at each

Table 1: Pedigree and identification of the four wheat genotypes used.

Genotype	Pedigree	Maturity group	Origin and specific features
KSU 102	Yecora Rojo x local varity sama	Medium	King Saud Univeristy, Faculty of Food and Agriculture Science, Plant Production Program
KSU 105	RI 474 x HD2172	Medium	King Saud Univeristy, Faculty of Food and Agriculture Science, Plant Production Program
Yecora Rojo	Cultivar variety	Semidwarf early mature	USA
West bread	Cultivar variety	Late mature	USA

Table 2: Number of irrigation and amount of water used for each treatment over the growing season (means of the two growing seasons of 2003/2004 and 2004/2005).

Water supply treatments	Mean of water apply over growing season (m^3/ha)	Mean of number of irrigations over growing season
Weekly irrigation	10000	20
Irrigation at 50 CPE	8000	16
Irrigation at 100 CPE	6000	12
Irrigation at 150 CPE	4000	8
Irrigation at 200 CPE	2000	4

irrigation (this depth of irrigation water is good enough to bring the soil to its field capacity). The CPE was calculated as a sum of daily-recorded evaporation from USWB class A open pan. Irrigation schedules were compared with the control (weekly irrigated) as a traditional method used by many farmers in the area of the experiment. At the end of the experiments the total amount of water supplied was calculated and is presented in Table 2.

2.3 Field experiments

Two field experiments were conducted during the winter seasons of 2003/2004 and 2004/2005 at the Agricultural Research Station, Derab, near Riyadh, Saudi Arabia (24^0 42' N latitude and 46^0 44' E longitudes, Alt 600 m). The climate in this region of the Saudi Arabia Kingdom has been classified as an arid climate. Before commencement of the field experiments, samples of the soil sites were taken for physical and chemical analyses according to the methods described by Cottenie et al [21] and But [22]. The results revealed that the soil sites were sandy-clay-loam texture (65% sand, 13% silt, 22% clay) with EC 1.4 dS/ m^{-1}. Water irrigation was also analyzed according to the methods described by American Public Health Association (APHA) [30]; the soil sites were prepared as recommended and divided into subplots. Each subplot consisted of eight rows, 2 m long and 20 cm apart. The total plot area was 3.2 m^2. The rate of 70 kg (P_2O_5) /ha in the form of super phosphate (16 % P_2O_5) and potassium fertilizer, in the form of potassium sulphate (42 % K_2O) by the rate of 100 kg/ha K_2O, were applied broadcasting during soil preparation, whereas the recommended dose of N (100 kg N/ha) was also applied in three split equal doses in the form of ammonium nitrate (33.3 % N) at sowing, during tillering and at anthesis. Seeds of each genotype were sown at a rate of 140 kg/ha by the hand drilled method on December 10 and 14 in the first and second season, respectively. With the exception of the experimental treatments, all agriculture practices for sowing wheat were carried out according to the conventional production practices followed in the Riyadh area. The irrigation system was a flooding irrigation system through line pipes provided by meter gages for measuring the water applied.

At harvest time, the inner four rows of each subplot unit were harvested to estimate yield and yield component characteristics. Water use efficiency (WUE) was calculated based on biological yield, according to the formula suggested by Bos [23] as follows:

WUE_b= above ground biomass as (kg) divided by seasonal water used in ET. ET is the amount of water used by the plants over the growing season in m^3. The harvest index (HI) and crop index (CI) were also calculated using the formula suggested by Donald and Humblin [24].

2.4 Statistical analysis

Data for each season were statistically analyzed according to the methods described by Gomez and Gomez [25]. The means were compared using least the significant difference test (LSD). However, the data obtained showed no significant differences between the two seasons of the study, therefore a combined analysis over the two growing seasons was done.

2.5 Economic evaluation

The economic evaluation of the data obtained herein was done based on the cost of inputs and outputs that prevailed during the period of experimentation as follows.

The gross return was worked out by multiplying the economic produce and by-produce (straw, haulm etc.) by respective sale prices and expressed in SR ha^{-1}. The net return for the data obtained was worked out by subtracting the total cost of cultivation from the total gross returns and expressed in SR ha^{-1}. The return SR^{-1} invested was calculated by dividing the gross returns by the total cost of cultivation.

3 Results

The data presented in Tables 3 and 4 manifests the effect of water irrigation treatments at different cumulative pan evaporation, CPE on grain yield and yield component characters of the four wheat genotypes. It is clearly obvious that low water supply treatments reduce all yield component characters under investigation. The data given elucidates that tillers number/m^2 was significantly reduced as irrigation supplies decreased. The highest number of tillers/m^2 (583.7) was recorded at 50 CPE, whereas the lowest number (499.4) was obtained when water irrigation was supplied at 200 CPE. The same picture was also observed in plant height: gradual decrease was in line with decreasing cumulative pan

Table 3: Effect of water irrigation levels on some yield and yield component characters of wheat genotypes (mean of two growing seasons of 2003 /2004 and 2004/2005).

Irrigation treat. at CPE (mm)	Mean of water applied m^3/ha	Plant height cm	Tillers number/m^2	Grain yield ton/ha	Decreasing %	Water conserving	
						m^3/ha	%
Control	10000	92.1	536.8	7.04	-----	-----	---
50	8000	93.1	583.7	6.87	2.41	2000	20
100	6000	77.1	517.5	5.97	15.20	4000	40
150	4000	79.0	510.6	4.99	26.12	6000	60
200	2000	73.6	499.4	4.48	36.36	8000	80
LSD $_{(0.05)}$		5.2	46.8	0.60	------	----	---

Table 4: Effect of water irrigation levels on some yield component parameters of wheat genotypes (mean of two growing seasons of 2003 /2004 and 2004/2005).

Irrigation treat. at CPE (mm)	Mean of water applied m^3/ha	Straw yield ton/ha	Biological Yield ton/ha	WUE_b Kg/m^3/ha	Filling period (day)	CI %
Control	10000	13.6	19.8	0.70	56.2	55.2
50	8000	12.8	18.3	0.86	57.3	60.1
100	6000	10.5	15.8	1.00	55.3	60.7
150	4000	10.8	14.0	1.25	53.6	55.4
200	2000	9.5	12.8	2.24	52.3	53.8
LSD $_{(0.05)}$		1.5	1.40	0.22	3.8	----

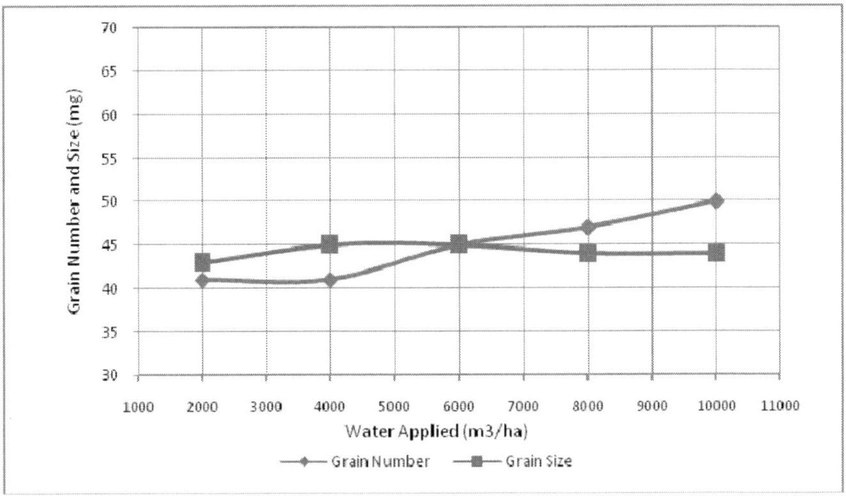

Figure 1: Grain number and size (mg) as effected by irrigation levels during the two growing seasons (2003/2004 and 2004/2005).

evaporation. The decrement was 77.1, 79.0 and 73.6 accompanied with irrigation treatments of 100, 150 and 200 CPE, respectively, as compared to the highest value. Noticeably, plants given the highest irrigation level (control) recorded lower values of tillers number/m^2, plant height, filling period and WUE as compared to irrigation water supplies at 50 cumulative pan evaporation. In addition, as expected grain yield was significantly decreased with decreasing

Table 5: Effect of genotypic variations on some growth characters of wheat genotypes (mean of the two growing seasons of 2003/2004 and 2004/2005).

Genotype	Plant height cm	Tillers number /m^2	Grain number/ Spike	Grain Size mg	Grain filling rate (g/m^2/day)
Yecora Rojo	70.28	590	34.4	47.7	7.6
West Bread	84.86	482	49.4	37.8	9.6
KSU 102	77.87	580	58.2	39.6	7.7
KSU 105	82.09	596	44.8	43.3	11.2
LSD $_{(0.05)}$	4.8	6.5	3.1	2.1	0.74

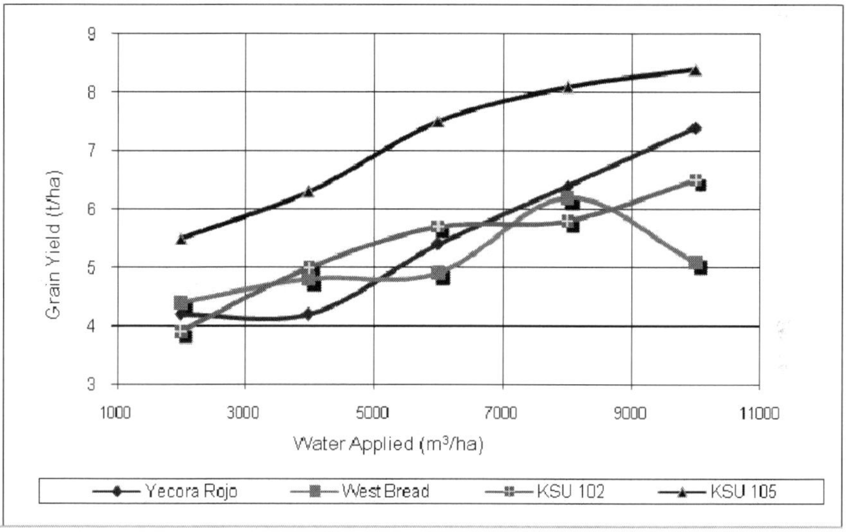

Figure 2: Yield productivity of four cultivars of wheat (t/ha) as effected by irrigation levels during the two growing seasons (2003/2004 and 2004/2005).

water irrigation levels irrespective of genotypes. With respect to grain yield (ton/ha), the results obtained indicated that the differences between the traditional method of water supply (7.04 ton/ha) and irrigation water applied at 50 mm of CPE (6.87 ton/ha) was not significant. Moreover, Fig 1 illustrates that grain number and grain size were affected by levels of irrigation supplies. Both characters were gradually decreased by increasing water deficit as a result of decreasing water irrigation level. Fig 2 also reveals that the tested four genotypes were different in their response to the water supply schedules. Promising genotype KSU105 surpassed the other genotypes and recorded the highest grain yield.

As regards genotypic variation, the data obtained demonstrated that advanced line KSU 105 outranked the other three genotypes and recorded the highest values of the most of studied characters under investigation viz., tillers number/m^2 (596), grain yield (7.14) ton/ha, biological yield (18.4) ton/ha and water use efficiency, Kg/m^3/ha as compared to the lowest values of the same attributed parameters were recorded by the West Bread variety viz., tillers number/m^2 (482), grain yield ton/ha (5.06), biological yield ton/ha (13.9) and water use efficiency, Kg/m^3/ha (Tables 5 and 6).

Concerning the effect of interactions between the two factors under investigation, i.e., genotypes x water irrigation schedule supply, the data obtained clearly indicated that by sowing line KSU105 in the presence of the highest level of irrigation water (10000 m^3/ha), the highest grain yield was obtained (8.41 ton/ha). Moreover, the data reveal that no significant differences between these treatments and a water irrigation at the level of 8000 m^3/ha, which recorded grain yield 8.08 ton/ha (Table 7).

With regard to economic evaluation, the results obtained in the present investigation, presented in (Table 8), reveal that the highest net return (2082.69 SR) and highest invested return per SR (1.37) were obtained by the irrigation of wheat plants under an arid environment at 50 followed by 100 CPE as compared to the other irrigation treatments.

4 Discussion

Drought is a global problem that is reducing plant growth and productivity worldwide; some of the most severe water deficits in root zones occur in arid and

Table 6: Means of yield and yield component characteristics of wheat genotypes as affected by genotypic variations (mean of the two growing seasons of 2003/2004 and 2004/2005).

Genotype	Biological Yield ton/ha	$WUE_{(g)}$ (Kg/m^3/ha)	Straw yield ton/ha	Grain yield ton/ha	HI %	CI %
Yecora Rojo	16.7	1.11	8.40	5.50	39.6	65.5
West Bread	13.9	1.09	11.64	5.06	30.3	43.5
KSU 102	16.2	1.11	10.81	5.39	33.3	49.9
KSU 105	18.4	1.48	11.26	7.14	38.8	63.4
LSD $_{(0.05)}$	0.8	0.12	0.42	0.40	------	------

Table 7: Means of interaction effect between irrigation levels and genotypic variation on grain yield of wheat genotypes (means of the two growing seasons of 2003/2004 and 2004/2005).

Irrigation treat. at CPE (mm)	Mean water applied m^3/ha	Yecora Rojo ton/ha	West Bread ton/ha	KSU102 ton/ha	KSU105 ton/ha
Control	10000	7.34	5.13	6.53	8.41
50	8000	6.38	6.15	5.77	8.08
100	6000	5.43	4.92	5.72	7.44
150	4000	4.17	4.75	5.02	6.31
200	2000	4.19	4.36	3.92	5.47
LSD $_{(0.05)}$		0.89	0.89	0.89	0.89

Table 8: Economic evaluation of the effect of irrigation schedules on some wheat genotypes grown under the arid environment of Saudi Arabia (mean of the two growing seasons of 2003/2004 and 2004/2005).

Irrigation treat. at CPE (mm)	Mean of water applied M^3/ha	Gross return (SR)	Total costs (SR)	Net return (SR)	Return per SR invested
Control	10000	7898.18	7019.2	879.0	1.13
50	8000	7701.89	5619.2	2082.69	1.37
100	6000	6692.91	5419.2	1273.71	1.24
150	4000	5594.24	5219.2	375.04	1.07
200	2000	5022.48	5019.2	3.28	1.00

SR: Saudi Riyal (1 $ USA = 3.75 SR).

semiarid regions of the world, where limited rainfall, high evapo-transpiration and high temperature play a very important role in leaf water content and plant water required. To solve the problem of drought two strategies could be used: provide plants with water by irrigation and/or selecting drought-tolerant genotypes. However, the cost and availability of irrigation water under arid and semiarid conditions make the irrigation approach unfeasible on a large scale. In respect of water supplies, previous results indicated that significant reduction in most of the yield and yield component characteristics was noticed due to decreasing water supply levels (Tables 3 and 4). Such effect of water supply may be due to the availability of water in the root zone promoted growth and proliferation of the root and thereby increased the absorption of water and soluble nutrients, resulting in a higher number of leaves per plant and large leaves, leading to increased LA. All of these parameters put together increase the rate of photosynthesis; better translocation of photosynthesis from leaves and stems to the sink finally favourably influenced the yield and yield component characteristics. Similar results were also reported in [1–6, 11]. In addition, the same trend was discussed by many authors [7–10, 12–14, 27, 28] who reported that low soil moisture content caused an irreversible loss in yield potential.

In the present study, yield and yield components were showed to be significantly affected by varietal differences. Among the cultivars used, line KSU-105 produced the highest grain yield (7.14 ton/ha) followed by Yecora Rojo (5.50 ton/ha), KSU-102(5.39 ton/ha) and West Bread (5.06 ton/ha). The superiority of line KSU- 105 may be because the selection of line KSU-105 was done at the same location as the present experiment. Thus, it has higher adaptability genetic constituencies toward drought conditions than the other tested genotypes. These results are in general agreement with those obtained in [15–20, 26, 29]. In the respect of water supply and varietal difference interactions, the data obtained in the present study (Table 6) indicate the importance of proper genotypes, which corroborated the findings of Ashour and Selim [28], who concluded that the effect of genotypic variations is one of the most important strategies for improving crop production.

5 Conclusion

According to the results obtained herein, it can be concluded that sustainable irrigation management under our arid regions can be achieved through selecting the most efficient irrigation schedule methods, such as CPE, where much of the water irrigation supply can be conserved in addition to adapting highly drought tolerant genotypes to achieve greater crop productivity and to reach sustainability under the stress conditions of arid and semiarid regions.

References

[1] Salter, P.J. and Goode, J.E., Crop response to water at different stages of growth. Res. Rev., common Bur-of Hortic. Plant Crops, No.2, 16, 1976

[2] Misra, R.D., Charma, K.C., Wright, B.C. and Sing, V.P., Critical stages in irrigation and irrigation requirements of wheat variety. Indian J. Agric. Sci. 39(9), 898, 1969.

[3] Chauhan, D.S., Hukkeri, S.B. and Dastance, N.G., Intensive versus extensive irrigation. Indian J. of Agron. 15(1), 46, 1970.

[4] Abd El-Ghany, H.M., Wheat production under water-limited sandy soil conditions using bio-organic fertilizer system. Egypt. J. Agron. Vol.29, No.1, 17-27, 2007.

[5] Hussain, G. and Al-Jaloud, A., Effect of irrigation and nitrogen on water use efficiency of wheat in Saudi Arabia. Agric. Water Manage., 27: 143-153, 1995.

[6] Ghandorah, M.O., El-Shawaf, I.I., Moustafa, Kh.A. and Gadallah, A.M., Evaluation of some early generations of bread wheat genotypes grown under heat and water stress at the central region of Saudi Arabia. Arab. Gulf J. Science. Res., 15(2): 505-523, 1997.

[7] Kramer, P.J. and Boyer, J.S., Water relation of Plant and Soils. Academic Press, San Diego, USA. pp4 95, 1995.

[8] Alderfasi, A.A., Ghandorah M.O. and. Mustafa, Kh.A., Evaluation of some wheat genotypes under drought stress in arid region of Saudi Arab Alex. J. Agric. Res. 44(3): 209-213, 1999.

[9] Reginato, R.J., Field quantification of crop water stress. ASAE, 26: 772-775, 1993.

[10] Reginato, R.J. and Carrot D.J., Irrigation scheduling with the crop water stress index. (C.F.springerLink- www.springlink.com), 1997.

[11] Imtiyaz, M., Mgadla, N.P. and Chepeze, B., Irrigation scheduling for vegetable and field crops. Irrigation Research paper 1, Department of Agriculture Research, Botswana, 36-42, 1994.

[12] Imtiyaz, M., Mgadla N.P., Chepeze B. and Manase, S.K., Response of six vegetable crop to irrigation schedules. Agriculture Water Management 45: 331-342, 2000.

[13] Pawar, J.K., Dhomane, A.S. and More, V.D., Studies on scheduling of irrigation and nitrogen levels on yield of summer sunflower. J. Indian Water. Res. Soc. 11(1): 57-59, 1991.

[14] Singh, S., Ram, M., Ram, D., Sharns, S. and Singh, D.V., Water requirement and productivity of palmarosa on sandy loam soil under sub-tropical climate. Agric. Water. Manag. 35: 1-10, 1997.

[15] El-Haddad, E.H., Amer, M.A. and Moistafa, M.A., Effect of salinity on the growth and yield of three wheat cultivars grown in calcareous soil. Menofiya J. Agric. Res. 18(3), 1669, 1993.

[16] Yousef, K.M.R. and Hanna, N.S., Evaluation of some bread wheat varieties to yield and water relations at Fayoum Governorate. Fayoum J. Agric. Res. & Dev. 12(2), 216, 1998.

[17] El-Beially, I.E.M.A, Effect of irrigation with saline water on growth, yield and its components of some wheat cultivars in sandy soil. Al-Azhar J. Agric. Res. 34, 59, 2001.

[18] Hassan, A.I., Moselhy, N.M.M. and Abd El-Maboud, MSh, Evaluation of some wheat cultivars under two levels of irrigation water salinity in calcareous soils, South Sinai. Zagazig J. Agric. Res. 29(1), 1, 2002.

[19] Abd Allah, A.E., Moustafa, M.A. and Salem, M.A., Interactive effect of saline irrigation water and zinc application on growth and yield of seven wheat cultivars grown in calcareous soil. J. Adv. Agric. Res. 8(3), 551, 2003.

[20] Reiad, M. Sh., Yasein, M., Tolba, A.M. and Abd-El-Samie, F.S., Effect of wheat genotypes, drought and soil salinity on growth, yield and its components. Egypt J. Agron Vol.29, No.1, pp.69-83, 2007.

[21] Cottenie, A. M., Verlo, L., Kjekens and Camerlynch, R., Chemical Analysis of Plant and Soil. Laboratory of Analytical Agro chemistry. State Univ. Gent, Belgium. No.42, pp.280-284, 1982.

[22] But, R., Soil Survey laboratory Manual Report No.42 USDA, National Resources Conservation Service .Washington, 2004.

[23] Bos, M.G., Summary of ICID definition of irrigation efficiency, ICID Bull, 34, 28-31, 1985.

[24] Donald, C.N. and Humblin J., The biological yield and harvest index of cereals as agronomic and plant breeding criteria. Adv. Agron., 28: 361-405, 1976.

[25] Gomez, K. A. and Gomez, A., Statistical Procedures for Agricultural Research 2nd ed. John Wiley & Sons, New York, 1984.

[26] Aderfasi, A.A., Response of four genotypes of wheat to irrigation schedules. Saudi. J. Biol. Sci., 7(2): 171-178, 2000.

[27] Aderfasi, A.A., Yield potential of two barley genotypes grown under water stress of arid ecosystem of Saudi Arabia, Am-Euras. J. Agric. & Environment. Sci.: 5(3): 348-353, 2009.

[28] Ashour, N.I. and Selim, M.M., Growth and Yield responses of some wheat cultivars to saline conditions in South Sinai Governorate. Egypt J. Agron. 19, No.1-2, pp. 139-148, 1994.

[29] Ghahrome, B. and Sepaskhah, A.R., Use of a water deficit sensitivity index for partial irrigation scheduling of wheat and barley. (C.F.springerLink www.springlink.com, 1998.

[30] American Public Health Association, Standard Methods for Examination of Water and Wastewater, 18th (ed) APHA, AWWA, WPCF, NY, Washington, 1992.

Robustness, vulnerability and adaptive capacity in social ecological systems: the case of spate irrigation systems in Pakistan

M. A. Kamran & G. P. Shivakoti
School of Environment, Resources and Development, Asian Institute of Technology, Thailand

Abstract

Socio-ecological systems (SESs), the spate irrigation system in our case, operate under the environment of stress from several external and contextual factors. The literature on collective action in Common Pool Resources (CPRs) considers non-storage and non-stationary characteristics of a resource as major obstacle to its sustainability. However, the case study systems have endured, despite acute scarcity and uncertainty of irrigation, for centuries in arid and semi-arid environments of the Middle East, Africa, South Asia, parts of South America and Central Asia, where rainfall is too low to support rainfed farming. These rainfall dependent irrigation systems are more complex and can exhibit unexpected changes as a result of climatic disturbances. Thus, actions directed at enhancing robustness to a particular set of disturbances can trigger changes in ecological dynamics that may, in turn, alter the set of disturbances faced by communities. The four case study systems for this research were selected from the Punjab (Pakistan) by purposive sampling, comprising two perennial and non-perennial farmer managed spate irrigation systems (FMIS) and two perennial and non-perennial agency managed irrigation systems (AMIS), providing 2X2 typology for analysis. The sustainability framework by Anderies et al. (*A framework to analyse the robustness of social-ecological systems from an institutional perspective*. Ecology and Society, 2004) has been used for analysis of this study.

Keywords: vulnerability, sustainability, adaptive capacity, irrigation management, arid and semi-arid environments, spate irrigation, Punjab Pakistan.

www.witpress.com, ISSN 1743-3541 (on-line)
doi:10.2495/SI100061

1 Introduction

Indigenous irrigation systems are facing new threats because of openness to the new world, the commercial interests of farmers, the rising cost of maintenance, increased competition for water and weakened social cohesion due to reasons that include state interventions [1]. Socio-ecological systems (SESs), irrigation systems in our case, are a complex collection of human, physical and institutional entities that respond to internal and external disturbances through a diverse array of rules in different conditions [2]. The nature of resources (mobile or stationary) responds differently to predictable and unpredictable disturbances [3]. The CPR theorists consider lack of storage and the non-stationary character of a resource, spate irrigation in this study, as a major obstacle for collective action [4]. The strong connections of SESs with larger scale phenomena pose challenges and opportunities for stakeholders. Literature based on past performance of resource systems shows that many long endured SESs have successfully adapted their institutions to these disturbance regimes [5], while others collapsed [6].

Spate irrigation is a type of water management system, which is unique to semi-arid environments, where flood water is generated by heavy rainfall in upper catchments and these systems use the occasional flow of floods to operate intermittently throughout the year [7]. The uncertainty about the occurrence of flood water, as well as the timing and size of the flood, are major factors leading to variation in cropped areas across cropping years [8]. Water rights in such systems are found to be complicated and conflicting, as different users have different rights depending upon the type of flow [9].

The spate irrigation systems have endured the known shocks of regular climate variability and floods with varying degrees of success in different management regimes and resource uncertainties [10], and are now exposed to climate related novel shocks and disturbances at a pace never experienced in human history [11]. Already, there is a 10–15% decrease in precipitation in arid areas of Pakistan and the predicted rainfall, particularly during the summer monsoon, is expected to increase [12]. Similarly larger drought periods, and the frequency as well as unfamiliar size and nature of floods, can have catastrophic impacts [13].

This paper looks into the adaptive capacity of spate irrigation systems, coping with past shocks in the form of appropriate institutions to manage external shocks (rainfall patterns, floods, droughts, etc.) and the system characteristics (run-off rivers diverted through indigenous structures); and estimating the chances of success of coping with novel shocks in the face of climate change. We have discussed our case study related issues, major threats and possible panaceas across the core components of a SES *viz* the resource system; governance system; resource units; resource users individually and in relationships that affect each other and the related ecosystems [14].

2 Methodology

In order to analyze the core entities of the SES and understand interactions between them, we adopted the framework proposed by Anderies et al. [15]. The framework focuses on four entities, which are mostly involved in CPRs harvested by people (Figure 1). The two entities in the framework namely, 'resource users' and 'public infrastructure providers' involve humans. The other two entities namely 'resource' and 'public infrastructure' involve physical and institutional aspects. The public infrastructure consists of human-made physical and institutional capital [16]. The 'resource' entity represents the biophysical system used by the 'resource users' through joint provision effort of the two human based entities in the framework *viz.* 'public infrastructure' and public infrastructure providers. The internal fluctuations can result from changes in relationships between resource users and infrastructure providers and can affect various components and linkages in the framework. Similarly, external threats can also affect components and interactions in the framework.

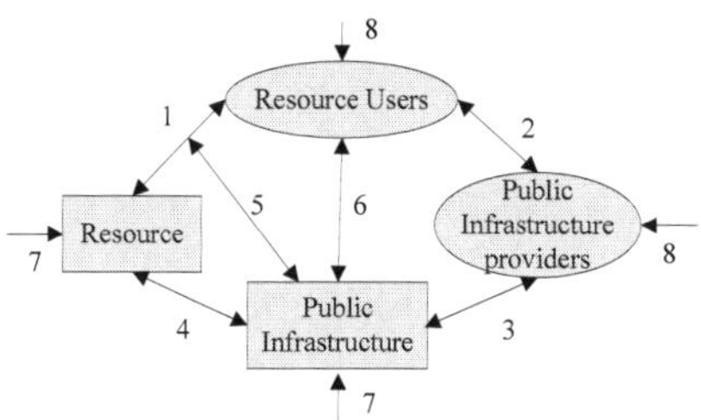

Figure 1: Anderies et al.'s [15] framework for SES sustainability.

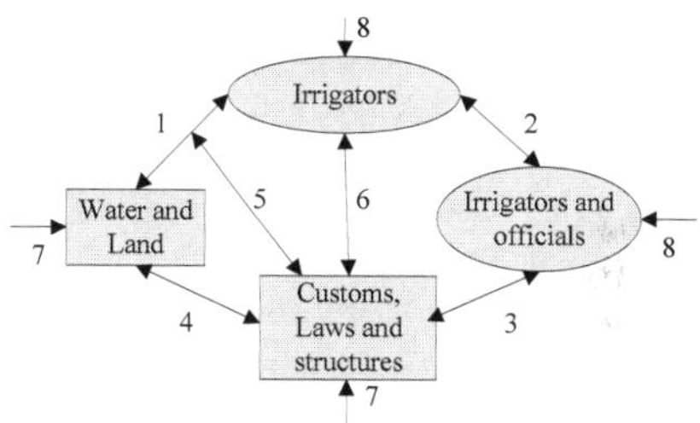

Figure 2: Modified framework for spate irrigation systems.

In FMIS, the resource users and public infrastructure providers are the same, and the factors that impact one entity also in return impact the other (if provision of public infrastructure is impacted by some factor, the users are also affected by it). In AMIS in our study, the human entities of the framework involve different actors i.e. public infrastructure is influenced by state departments and officials, with less influence of local communities compared to community managed systems.

3 Analysis and discussion

3.1 Institutional change in spate community irrigation systems

The deterioration of local resource management institutions by neglect of colonial powers is well documented in forestry, irrigation and fisheries resources [14]. Pakistan's water management policies are largely continuation of British colonial legacy [17]. These policies were made for consolidation of colonial control rather than concerns for public well-being and equity [18]. The irrigation control provided strong link between local communities and colonial state in such an arid environment [17]. The inability of the state to change colonial laws resulted in continued policing role of the state rather than that of a service provider [18]. The comparative analysis of "Riwajat-e-Abpashi (1864)" and "Minor Canal Act (1905)" shows that the colonial act transferred powers to state from communities in respect of Operation, Maintenance and management of spate irrigation systems and found that the act provides greater authority and decisions making powers to the state which is quite conflicting with the customary rights.

Now after the analysis of process and nature of institutional change, we move forward to understand how such changes lead toward management of these systems.

3.2 Cropping patterns in a changing context

The choice of crop(s) is highly influenced by availability of water and the cropping pattern varies with type of spate irrigation system and supplemental water availability. Usually, sorghum, millets and Chickpea are major summer crops and the next wheat crop cannot be cultivated in the fields with summer crops as there is not enough residual moisture to support it. Similarly if fields get subsequent spate water later in summer season (August-September) then the moisture will be stored and land will be allocated for wheat instead of summer crops (in this case the land is ploughed even if any other crop was sown in that field). In lowland perennial systems have adapted to wheat and cotton or rice cropping systems (with sorghum and vegetables as mixed crops). The rain-fed farming is practiced for growing fodder on non-haqooq or lands out of spate command. Due to growing scarcities with bringing more areas at head ends, clear distinction can be observed between head end and tail ends crops, crop

Table 1: Impact of institutional change on agricultural production systems.

Impacts on agricultural production systems	AMIS				FMIS			
	Perennial		Non-perennial		Perennial		Non-perennial	
	Head	Tail	Head	Tail	Head	Tail	Head	Tail
Customary Land use followed	12.9	21.4	32.9	40.0	82.9	90.0	81.4	88.6
Increase in number of crops	87.1	18.6	31.4	14.3	14.3	8.6	11.4	21.4
Decrease in number of crops	0.0	60.0	35.7	45.7	2.9	1.4	4.3	0.0
Inclusion of commercial crops	67.1	24.3	17.1	25.7	32.9	25.7	0.0	0.0

Note: Figures show % of yes response.

yields and cultivated area. Also the declining kamara system has weakened communities' ability to stop and or divert big floods; only small and medium floods can be used for irrigation most of which is taken by head end lands.

3.3 Eroding social capital and declining collective action

The water users in indigenous irrigation systems can have conflicts in various water management aspects. However, only few members talk about conflicts in such homogenous societies in order to maintain communal harmony [2]. Therefore, the respondents in this study were asked to name and prioritize problems associated with collective action for water management and related risks to their traditional irrigation. The institutional risk values were calculated by calculating and Incidence (I) and severity index (S) [19]. The severity index shows the severity of risk of a problem on a scale of 1 and 2 (where 1 shows 'most severe' and to 2 represents 'least severe') and the incidence index shows how many respondents reported that problem (with '1' representing incidence and '0' representing non-occurrence of problem in a response). The major benefit of this method was that the respondents were neither fixed with the number of problem to report nor they were fixed with the choices to select among. The severity index for a problem shows the order in which a respondent mentioned the problem in total responses. The mean value of the responses was then calculated for the respondents who reported a particular problem in survey.

Table 2: Perceived institutional risks.

Risks to institutional sustainability	Overall Index	AMIS		FMIS	
		Perennial	Non-perennial	Perennial	Non-perennial
Violation of rules	0.47	0.60	0.57	0.33	0.36
Officials intervention in customary laws	0.30	0.59	0.37	0	0
Declining relative value of farming	0.52	0.51	0.60	0.42	0.54
O&M cots	0.47	0.43	0.32	0.52	0.56
Outmigration	0.43	0.54	0.54	0.36	0.18
Decreasing land holdings	0.55	0.45	0.54	0.63	0.58
Support from government in losses due to very heavy floods	0.56	0.33	0.46	0.62	0.75
Extension and research services	0.41	0.07	0.31	0.51	0.56

The information was then used to construct a composite risk index to compare these issues across systems and across different management regimes.

The index values show that higher violation, interventions in customary laws and conflicts in operation and maintenance and management of the systems were viewed as high risk factors in AMIS compared with FMIS. However the decreasing land holding due to distribution (among all children according to Islamic law) and lack of government support even in very large floods which destroy crops and demolish structures was viewed as a major constraint by farmers in hilly FMIS. The high indexes for extension and research support in hilly area and non-perennial spate respondents shows that they consider lack of specific research and extension services for crop and livestock as major risk to sustainability of spate farming. The outmigration in AMIS was created problems in terms of abandoning farming in those lands and increased labour costs for remaining farmers (non-perennial systems mainly) and handing over lands for rental farming and new entrants prefer to grow non-traditional cash crops and high yielding varieties with the use of pesticides and fertilizers which create hazards for downstream drinking water for humans and animals (the case study FMIS perennial systems has traditional drinking water rights equal to two fingers width in the main lateral).

3.4 Performance of the systems and outcomes

The performance of irrigation systems can be measured by variety of methods including sophisticated technical ways using marginal productivity, irrigation efficiency, conveyance efficiency etc. Our objective is to compare the performance of systems which are different in their absolute performance

Table 3: Comparison of performance indicators in selected systems.

Performance measures	AMIS				FMIS			
	Perennial		Non-perennial		Perennial		Non-perennial	
	Head	Tail	Head	Tail	Head	Tail	Head	Tail
Participation in maintenance work	24.3	15.7	60.0	47.1	67.1	75.7	84.3	88.6
Conformance with water distribution rules	44.3	84.3	80.0	68.6	81.4	91.4	90.0	98.6
Water adequacy considering system characteristics in average years	78.6	38.6	67.1	50.0	87.1	95.7	77.1	70.0

Note: Figures show % of 'yes' response.

characteristics (like perennial and non-perennial can be measured using same scale). Therefore instead of going into absolute measurement methods, we analyzed systems performance using outcome criteria [20], like participation in maintenance; rules conformance and adequacy of water under different scenarios at head end and tail ends of the system (higher participation, better conformance to rules and adequacy compared with systems actual capacity show better performance of system and vice versa).

The external interventions in the form of institutional change impaired system performance in the form of participation in collective work and water adequacy and violence of rules. The external interventions in the form of institutional change impaired system performance in the form of participation in collective work and water adequacy and violence of rules. The results show that FMIS performed better than their corresponding AMIS. The major differences can be found in overall responses about system performance in general and difference in head end and tail end responses in case of AMIS. The situation is even worse in case of perennial AMIS where the higher productivity; potential to grow commercial crops, lower labour requirements coupled with officials ability to manipulate promoted opportunistic behaviour by head-end farmers.

3.5 Robustness, vulnerability and adaptive capacity of the systems

Our study is based on the hypothesis that the institutional change without giving due consideration to possible effects to other entities in SESs leads to counterproductive output. The analysis of changed role of resource users and public infrastructure providers (human components of the Anderies et al, 2004 framework) resulted in the form of variation in operational-level rules and variation in interventions in the form of structural improvements of systems. The interactions among the entities of the core subsystems of SESs (Table 4) shows difference in robustness due to variation in capacity to cope challenges associated with these systems. The inherent uncertainty in these systems has been a major threat and will be further aggravated due to climate change and preparedness of systems under two different management regimes with variation in resource uncertainty has been analyzed in detail.

The interventions by public infrastructure providers' in the in the form of changed rules at constitutional choice level impacted operational level rules through changed role of actors at collective choice arenas in AMIS. This further impacted the public infrastructure by construction work which resulted in decreased willingness of users in the maintenance work despite of high failure rates and lack of functioning of these structures after few huge floods. This change bring with it poor monitoring responsibility with government officials in the form of scanty staff of low pay scales and more authority, leading to corruption and rent seeking. This resulted in overriding further promoted by prolonged and expensive court cases.

On the other hand the FMIS continued to remain under same institutional structures with centuries old water use traditions and formulated appropriate rules to cope the external shocks. The rules in the form of reactive water rights,

Table 4: Linkages involved in SESs to cope with climate change and other challenges.

Linkages among core entities	Major Threats to spate systems	AMIS	FMIS
Between resource and resource users	Changing rainfall patterns resulting in variation in time, duration, and size of floods	Fixed rules implemented by state have exacerbated the situation of coping unusual climate events.	Reactive water rights have potential to cope climatic change induced rainfall patterns.
Between users and public infrastructure	Declining deliberation process and weak monitoring.	i. Conflicting state and community laws and officials dominates decisions ii. Officially appointed lower staff involve in corruption.	i. Irrigator's agreed customary laws which are reactive in nature ii. Locally selected monitors paid proportionate to total production makes irrigation fair.
Between public infrastructure providers and public infrastructure	i. Fixed maintenance plans ii. The increasing high floods events in frequency and intensity.	i. Meeting is called by official decree without consulting local farmers ii. High floods damage permanent state structures and sometime demolish them.	i. Farmers' gathers before rainy season and select maimar and or irrigators for the season/year ii. The low cost structures are rebuild using local material; once destructed by floods.
Between public infrastructure and resource	Concrete build engineering works fail to address flood challenges and customary laws.	Reliance on permanent structures made by short term projects affected community participation by realization that state should maintain infrastructure.	Only locally constructed earthen structures exist which are maintained and build every flood season by community and has kept collective action intact.
Between public infrastructure and resource dynamics	Saline ground water unfit for irrigation. Pumping of saline water to fulfil cash crops water requirements.	The farmers growing commercial crops and new varieties supplement with saline water which leads to salinity problem.	People following customary laws are following traditional land use and are not confronting any salinity issues.
Between resource users and public infrastructure	Conflicts, court cases, head-tail and free riding issues.	Tension among members and high costs as cases go to police and even at high court level in some instances.	Decisions are made at local level and most conflicts are mediated by tribal elders at community level.
External forces on resource	Changing flood paths with erosion of flood plain due to heavy floods.	The fixed rules and fixed points for water diversion creates problem in emerging external challenges.	The reactive water rights help farmers to adjust diversion points to changing flood paths.
External forces on resource users	Preference for non-farm jobs due to economic integration and decreasing returns due to climatic shocks	The outmigration has resulted in higher costs per unit of irrigated land for maintenance and operation.	The flexible water laws provided equitable distribution of water to all members and resulted in continued farming in these areas.

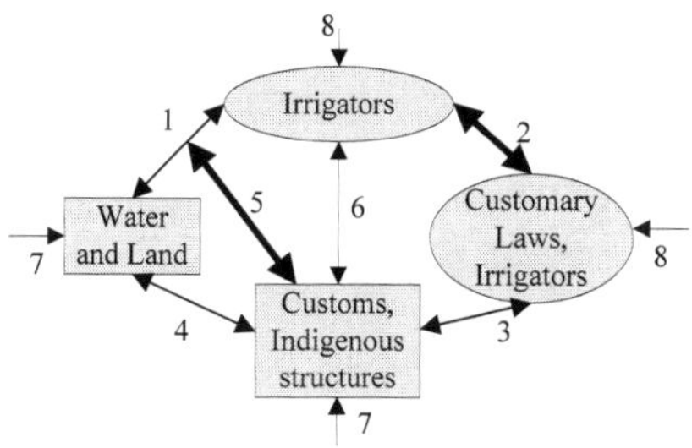

Figure 3: Modified framework for FMIS (bold lines represent strong linkages).

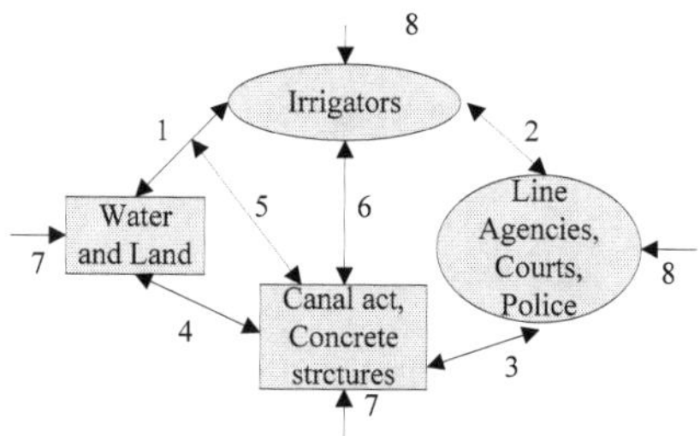

Figure 4: Modified framework for AMIS (dotted lines represent weaker linkages).

with more flexibility and fair community monitoring mechanism, kept the functioning intact and resulted in superior outcomes in the form of participation in the maintenance work and maintaining equity across the systems.

The major difference in the two management regimes was found in stronger relationship among resource users and public infrastructure providers in FMIS which resulted in robust public infrastructure and resource dynamics (Figure 3). On the other hand, in AMIS, the weakened relationship between resource users and public infrastructure providers resulted in weak public infrastructure and its fragile relationship with resource dynamics (Figure 4).

4 Conclusion

We have looked into dynamics of four spate irrigation systems in the southern Punjab part of Pakistan in the context of variation in governance and resource availability. In the plain areas systems, the state interference to get better hold on key resources in colonial times, transferred management roles to local landlords (in the form of *'Patti Daar' or 'zamindar')* and government officials, who continued with the customary laws until the enactment of the minor canal act of 1905. The state sponsored machinery and construction of permanent diversion and protection structures, reduced needs for labour and bullocks and provided chance to local influential farmers to use rules in their favour where officials were granted final authority to make any changes in irrigation rights and change in watercourses and canals. The results shows that collective action and local institutions are fragile and more vulnerable to external shocks in perennial systems where lower cost of maintenance compared with returns and higher interest of officials to manipulate rules to favour head-end influential landlords provided incentive to use rules against customary laws. This resulted in fixed rules, violation of customary laws, more conflicts, weakened authority of traditional water masters and growing conflicts and court cases.

A complete analysis of entities involved in the system was done to understand institutional issues in the systems [14]. It was given equal importance to understand how rules were designed and established and how far the users adapted to these rules and desire any change for better resource management. The institutional diversity provides better opportunity and greater social capital for institutional design [21]. The results shows that the FMIS proved to be more robust by adapting to shocks of changing economic, socio-political and ecological conditions through better appropriate institutional design which can also endure future shocks [15]. This study has also identified stronger and weaker linkages in the linkages among the core entities involved in SESs in the two management regimes. It is important to understand relationships among all entities involved in SESs, before making any policy to make systems more robust and sustainable institutions.

References

[1] Barker, R. and Molle, F. *Perspectives on Asian Irrigation.* In G. P. Shivakoti, D. Vermillion, W. F. Lam, E. Ostrom, U. Pradhan, and R. Yoder, eds. *Asian Irrigation in Transition: responding to the Challenges*, 2005.

[2] Shivakoti, G.P. and Bastakoti, R.C. *The robustness of Montane irrigation systems of Thailand in a dynamic human-water resource interface*. Journal of Institutional Economics, pp. 227-247, 2006.

[3] Janssen, M.A., J.M. Anderies and E. Ostrom. *Robustness of Social-Ecological Systems to Spatial and Temporal Disturbance Regimes*. Society and Natural Resources, pp. 307-322, 2007.

[4] Ostrom, E., Gardner, R. and Walker, J. *Rules, Games, and Common Pool-Resources*. The University of Michigan Press, 1994.

[5] Agrawal, A. *Greener pastures, Politics, Markets, and Community among a migrant pastoral people*, Durham, NC: Duke University Press. 1999

[6] Baker, J.M. *The Kuhls of Kangra*: *Community-managed irrigation in the Western Himalaya.* 2005

[7] Mehrai, A., Schulz, B., and Depeweg, H. *Hydraulic performance evaluation of the spate irrigation systems in Eriteria.* Irrigation and Drainage, pp. 1-14, 2005.

[8] Steenbergen. F.V. *Understanding the sociology of spate irrigation in Balochistan.* Journal of Arid Environments, pp. 349-365, 1997.

[9] Vincent, L. *Hill Irrigation Water and Development in mountain irrigation.* Intermediate Technology Publications. Overseas Development Institute. 1995

[10] Kamran, M.A. and Shivakoti, G.P. *Design principles and robustness of spate community irrigation systems in the Punjab Pakistan.* Workshop in political theory and policy analysis, Indiana University, 2009.

[11] Thornton, P. K. and Jones, P.G. Mapping climatic vulnerability and poverty in Africa. Report to the Department for International Development pp. 171, 2006.

[12] IPCC. *Climate Change 2007: Impacts, Adaptation and Vulnerability.* Contribution of Working Group II to the Fourth Assessment Report of the

Intergovernmental Panel on Climate Change, Published for the Intergovernmental Panel on Climate Change. Cambridge University Press, 2007

[13] IFAD. *Spate irrigation, livelihood improvement and adaptation to climate variability and change.* Commissioned by International Fund for Agricultural Development, 2009.

[14] Ostrom, E. *A general framework for analyzing sustainability of social-ecological systems*. Science, 2009.

[15] Anderies, J.M., Janssen, M.A., and Ostrom, E. *A framework to analyse the robustness of social-ecological systems from an institutional perspective.* Ecology and Society, 2004.

[16] Ostrom, E. and Ahn, T.K. *Foundations of Social Capital.* Cheltenham. U.K., 2003

[17] Gilmartin, D. *Empire and Imperial Science: Colonialism and Irrigation Technology in the Indus Basin.* The Journal of Asian Studies, pp. 1127-1149, 1994.

[18] Mustafa, D. *Colonial law, contemporary water issues in Pakistan.* Political Geography, pp. 817-837, 2001.

[19] Smith, K., Barrett, C.B., Box, P.W. *Participatory risk mapping for targeting research and assistance: with an example from east African pastoralists*, World Development, pp.1945–1959, 2000.

[20] Tang, S. *Institutions and collective action: Self-governance in irrigation.* Institute of contemporary studies press, San Francisco, 1992.

[21] Bloomquist, W. and Ostrom, E. *Deliberation, learning, and institutional change: the evolution of institutions in judicial settings.* Constitutional Political Economy, pp.180–202, 2008.

Five thousand years of irrigation history in Iran: rituals and methods

M. Sadrzadeh[1] & G. Badalians Gholikandi[2]
[1]*Faculty of Foreign Languages and Literature, Tehran University, Iran*
[2]*Power and Water University of Technology (PWUT), Water Research Institute (WRI), Iran*

Abstract

Iran's geographical location and the low availability of water resources are the reasons for water receiving a special sanctum. Water has a mythic root in Iranian culture that has been used in tales and stories. The first official and documented knowledge of irrigation, water sharing and its use in ancient Iran dates back more than five thousand years. Ancient structures and customs that have been deeply rooted in Iran's history can still be found in several parts of the country. In this research, special customs such as invokes for the start of or the end of rain (sun invoke), as well as flood irrigation and constructed structures over the last 2500 years, were surveyed in central parts of Iran. In regard to water distribution systems and their importance in dry and semi-dry areas, the results of irrigation system of ancient Drakan city and its countryside in Fars province of 10-13 A.D. were studied.

One of the conventional water resources management systems in ancient Iran was the water intake from saline rivers in specific seasons of the year, and several weirs, intakes and channels were constructed. In this study, survey results of the Abshour River that was located beside Fadami plain, Darab, on today's border of Fars and Hormozgan provinces, and constructed structures for the irrigation of local plains 1300–1500 years ago are given.

Keywords: water, ancient Iran, irrigation, Fars, Drakan.

1 Introduction

Iran's geographical location and low available water resources have caused Iranians to search widely for water, which also has had a special value and

www.witpress.com, ISSN 1743-3541 (on-line)
doi:10.2495/SI100071

sanctity. Iranians always know the water as a source of godsend and welfare, so they have special customs for worship and thanks.

Water has been so important that Iranians have called their provenance "water and land" [1]. Rainfall invokes, or those for the end of rain, have been some of the peoples' beliefs about water. Many proverbs and phrases about water can be found in Iranian culture. Water has a mythic root in Iranian culture. According to the historical documents and also Ferdosi's Shahnameh, Houshang (the king of Achaemenid dynasty) taught people irrigation methods, Deim and water farming, and wheat reaping. Hence, the first document about the irrigation history of Iran is from about five thousand years ago [2].

The ancient Iranians have been known not only for their science and technology in water structures, but also they were famous in the management and organization of water resources. In the Sasanian era, and years after that, an integrated rule, Namak, was written and applied, and a "Divan kast fozud" organization acted on the bases of that rule. This organization was a powerful organization that controlled the construction of water structures and their operations and maintenance. Moreover, this organization resolved water sharing issues between people, determining and registering water shares based on the Namak rule [3].

2 Water uses for irrigation and mill operation

In ancient Iran, water transfer from springs to the point of use and proper sharing were always important (for example the "Drakan" city in Fars province).

In the remaining geographical documents of 700-1000 years ago, a city was named Drakan or Darkan which is omitted in today's maps. Today, a city ruined by an earthquake is located 2 km to the southeast of the Darakou village (Tal Narenjoon) and researchers believe that that city is Drakan city.

In the debris of the Drakan city, many signs of stony and concrete canals and tanks were found. In addition, water structures in the city and country reveals that water was transferred from long distances. First, water storage tanks were filled, and then water was transferred to the farms. Some mills were constructed and installed in the flow of water canals to utilise the application of water energy.

Drakan city is located in a situation surrounded by farming lands, so several water structures should be constructed in that valley for proper water supply to farms. Abdalan, Abareh, Savareh, Canals, Tondabeh and Siphons are such of those structures. The required drinking water of inhabitants were supplied from mountainous paths by stony and concrete canals and stored in storage tanks. The remarkable point in Drakan city is that inhabitants weren't losing excess water and were conducting excess water to their farms [4].

3 Water transfer structures

There are many mountains and valleys in extent parts of Iran. Research shows that different water transfer structures have been applied since three thousand

Figure 1: The Makhak pond.

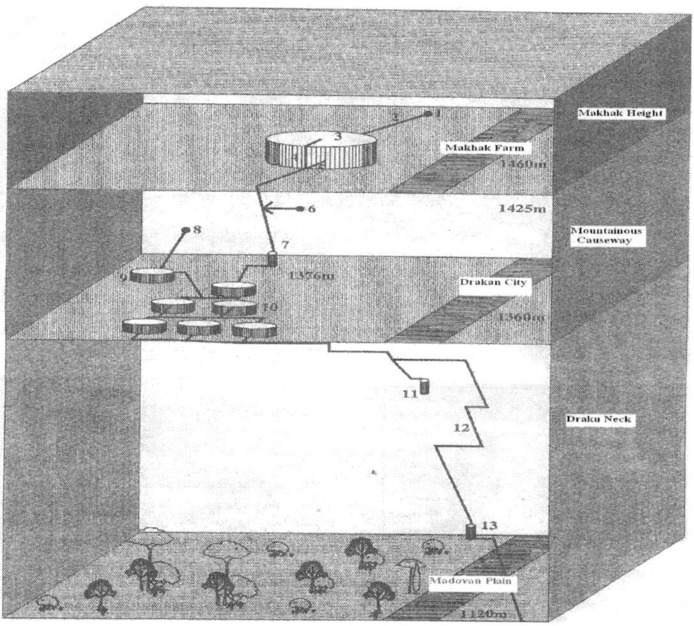

Figure 2: Water distribution from Makhak farm (1460 m) to Drakan city (1360 m) and Madovan plain (1320 m). 1- Jarm Sangi spring, 2- water canal, 3- Makhak pond, 4- weir, 5- hatch, 6- canal 7- mill 8- canal, 9- canal pond, 10- Drakan ponds, 11- mill, 12- water canal, 13- mill.

Figure 3: Water distribution canal from Makhak to Drakan city.

Figure 4: Canal stucco for prevention from water infiltration.

Figure 5: Orifices in water canals for Madovan farms irrigation.

(a)

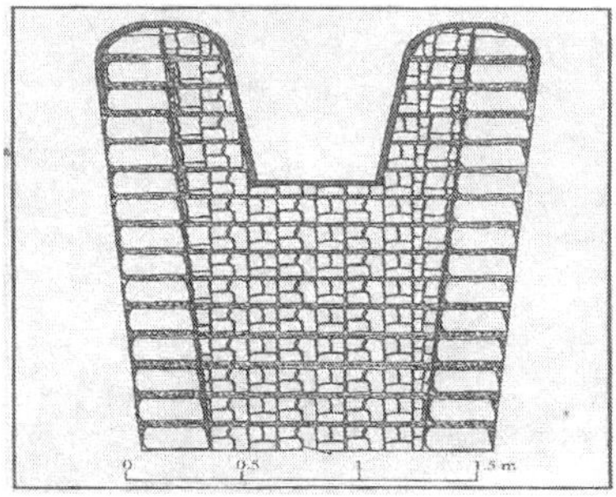

(b)

Figure 6: Ruined weir on Darakou valley stream: (a) direct view; (b) cross-sectional view.

(a)

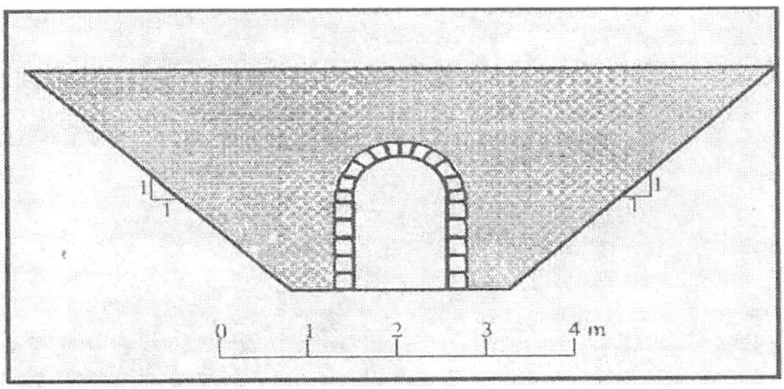

(b)

Figure 7: A water distribution weir from Drakan city to Madovan plain: (a) direct view; (b) cross-sectional view.

Figure 8: Two ponds at Dokhtar Darakou castle.

Figure 9: Scheme of the twin mill of Masjed Sangi.

Figure 10: Quadrilateral shaped canal carved in a stone, Dehrood river.

years ago. One of the most popular is Jouy, which is named in English as a canal. The canal word is from the Ariaian language. In the Ariaian language, the word “Cal” was used in names of many streams and rivers [4].

4 Rain invoking

One of the similar aspects of most human customs is interest in the sky and its phenomenon. In general, rain invoking customs are more common in arid and semi-arid areas. In these areas, farming acts have been based on rain, especially in the spring season. The rain invoking costume has been important in most parts of Iran, especially in arid and semi-arid areas. Many reports about this field are related to Khorasan, Isfahan, Fars and Khouzestan provinces. The rain invoking custom has not been documented and similar to many other customs, has been

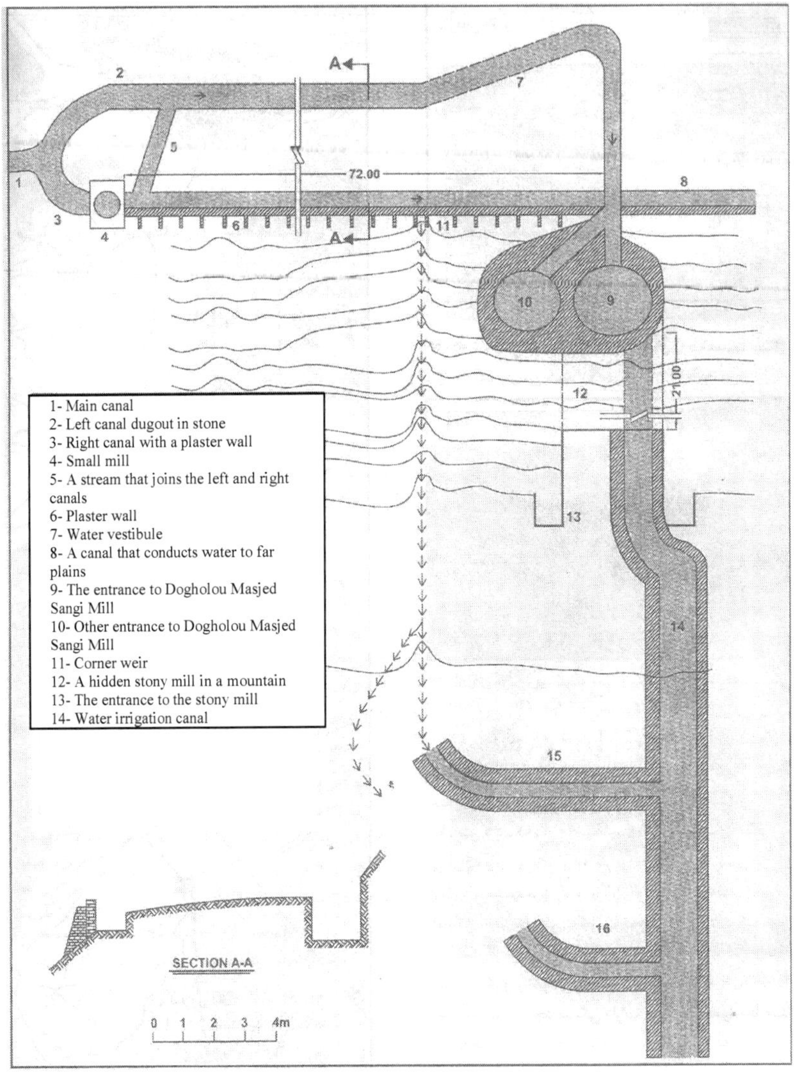

Figure 11: Several sections of the twin mill at Masjed Sangi in Darab.

unwritten, thus many changes have been entered in these customs. In ancient Iran, from the Hakemenenian era, the 13th day of each month was a Rain Goddess day (Tashtar day). In the New Year, Iranians go out to the environment and in the midday that Malvolent Apoush (the black monster of drought) is set back by Tashtar, sacrifice a sheep to have a rainy and talented year. The ancient Iranians knew of "Anahita" as a water goddess and painted and worshipped her picture that had been in each shrine (Figure 15).

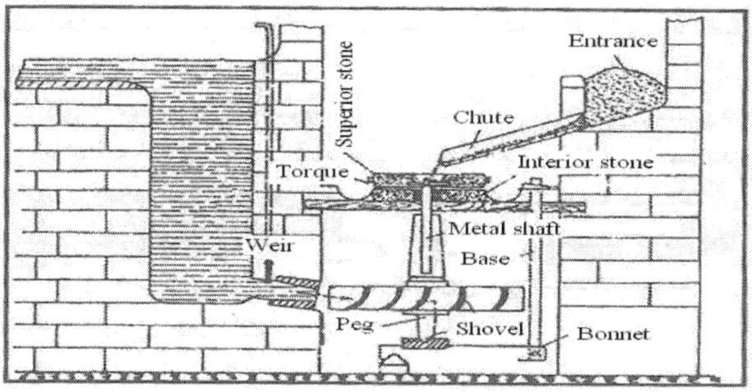

(a)

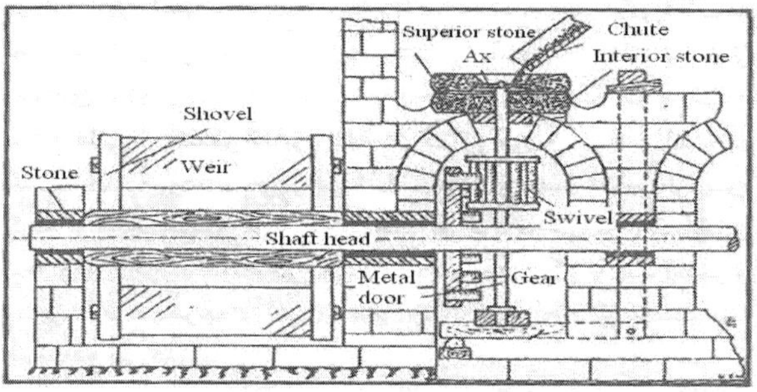

(b)

Figure 12: Ancient mills: (a) chute mill; (b) wheel mill.

In this custom, a manikin which may be the Rain Angel symbol, had several names in different areas or a man with long and postiche beard or a beardless man (drought symbol) was used and some poems and lampoons were sung by youths about them. Sometimes people with their babies went out into the plains to invoke for rain. The most interesting aspect of this custom is water pouring on the manikin or each other. By Islam propagation in Iran, the rain invoking custom was changed to rain prayers in which Muslims were praying and fasting for acceptance of their invoke for rain. In ancient Iran, the rain invoking custom had a long history and a praying field was in each city for rain praying.

The rain angel in the water pouring custom is changed to a Sayyid woman after the introduction of Islam to some villages of Iran. However, this custom in more developed cities of Iran is changed to a Sayyid man, who prays for the rain. Moreover, in some parts of Iran the manikin or cooking the Balghour soup have been substituted by charity gatherings of people and cooking foods for inhabitants in a mosque [1].

5 Conclusion

In Iran, many parts are subjected to drought and have irregular rain periods. In the past, this irregularity caused many droughts and then heavy floods in rainy

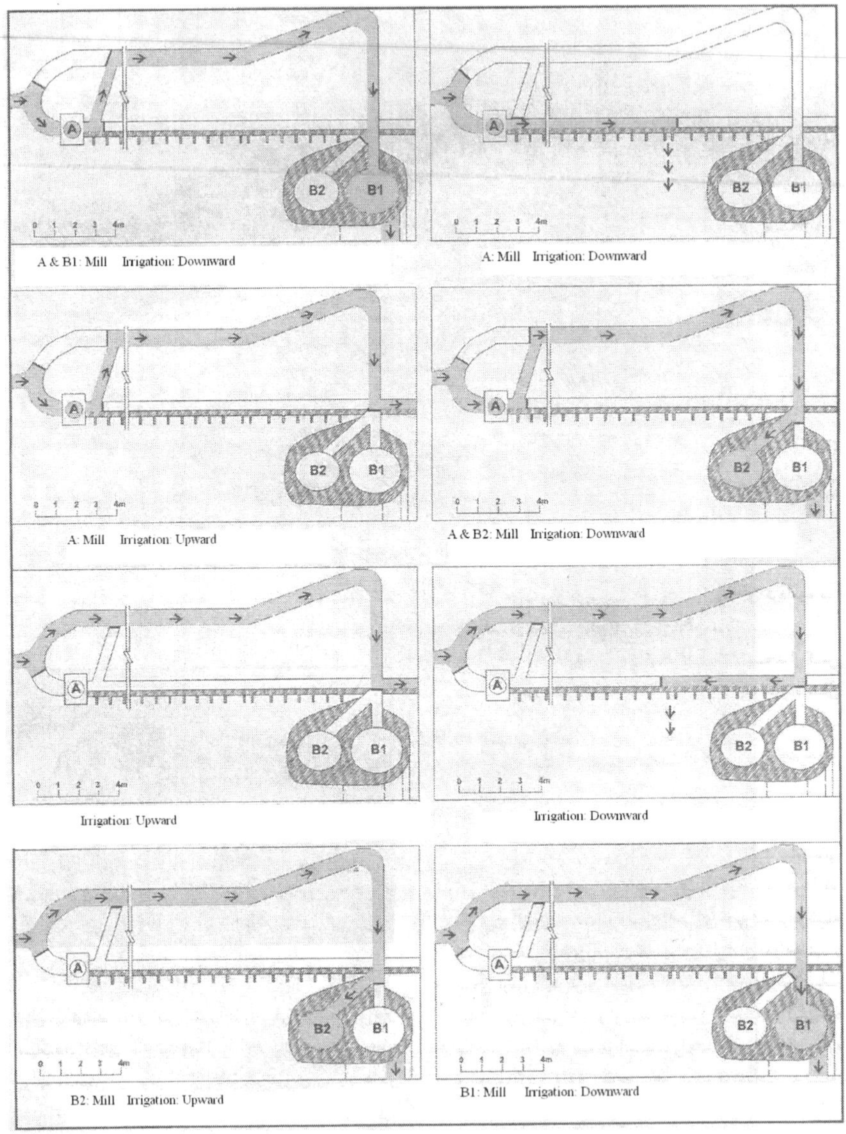

Figure 13: Several operating conditions of the Masjed Sangi Mill and irrigation.

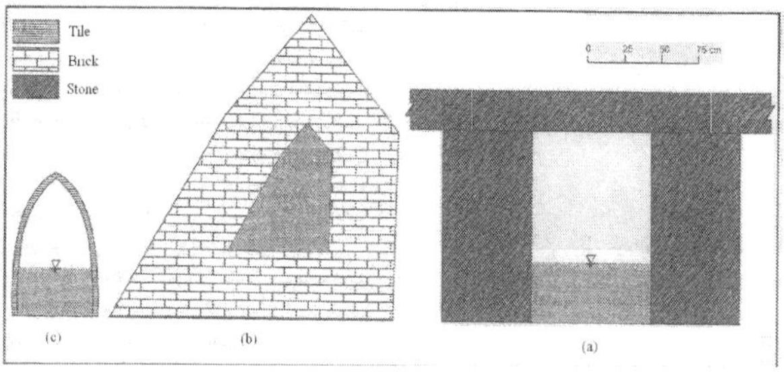

Figure 14: Different porch canals of ancient dynasties: (a) Achaemenid; (b) Zand; (c) Safavi.

Figure 15: The Anahita goddess, Sasanian dynasty.

periods. Hence, finding a resolution for these problems by Iranians was inevitable. Iranians always solve natural disasters and also profit from them; store water in ground and underground, transfer water to different parts, applying several irrigation methods and grow plants which are endurable with special climates.

In ancient Iran, people always used applied engineering methods and best natural and manmade stuff for implementation of weirs, cisterns, mills and other water structures. Therefore, Iranians could establish powerful and vast emperors like Achaemenian and Sasanian dynasties, in this drought land. These emperors established as a result of providing infrastructures and laws for applying them. Moreover, water in Iran has had a holy and worthy value from past to now and many customs existence in Iran's culture prove this claim.

References

[1] Khalatbary, M. (2008). *Water, its customs and beliefs in the national culture*. Soroush Publication, Tehran.

[2] Safinejad, J. (1997). *Water and irrigation history in Iran.* In proceeding of Iran's water and irrigation conference, The irrigation and drainage committee, Shiraz.
[3] Oyoun Al-Akhbar. (1948). *Water and irrigation technology in ancient Iran.* Cairo, Vol. 1, 44.
[4] Javaheri, P., Javaheri, M. (2006). *Water crisis in the Persian Gulf.* The regional water company of Fars, Boushehr and Kohkiloye& Boyer Ahmad.

Section 2
Modelling and simulation

Decision models for better land management under cultivation

M. C. Cunha[1], T. Leitão[2], J. P. Lobo-Ferreira[2] & L. S. G. Oliveira[2]
[1]*Department of Civil Engineering, University of Coimbra, IMAR, Portugal*
[2]*LNEC, National Laboratory for Civil Engineering, Portugal*

Abstract

Integrated water resources management is an issue of major concern in Europe. Preoccupation with environmental matters has been dramatically increased by the growing impact of various activities, such as farming. In fact, diffuse pollution provoked by agricultural practices is among the environmental challenges for which preventive measures are needed to achieve a "good status" regarding ecological and chemical parameters. Building models to support future decisions on better policies for rural land use planning (type of crops and associated fertilizers and treatment techniques) is crucial for achieving a sustainable development. This paper presents two decision models that aim to minimize diffuse pollution induced by agricultural land use and irrigation practices.

Keywords: diffusion pollution, sustainable agriculture practices, better land use decision models.

1 Introduction

Diffuse pollution produced by irrigation has become a major environmental problem and much research is being done into better land use (type of crops and their fertilizers, treatment techniques and irrigation practices), taking into consideration the vulnerability and risk concepts. Multidisciplinary approaches are needed if all the issues involved in the integrated management of land and water resources are to be represented. The decisions to be implemented should be carefully defined in order to achieve sustainable agricultural land use and production practices. The Water Framework Directive establishes a legal

WIT Transactions on Ecology and the Environment, Vol 134, © 2010 WIT Press
www.witpress.com, ISSN 1743-3541 (on-line)
doi:10.2495/SI100081

framework to protect surface water and groundwater use; therefore irrigation practices face new challenges. In fact, stating that "Member States shall implement the measures necessary to prevent or limit the input of pollutants into groundwater and to prevent the deterioration of the status of all bodies of groundwater" means that a new framework must be established for irrigation planning and management procedures. Decision models can significantly contribute to the design of best irrigation and land use practices (Kumar *et al.* [2], Manocchi and Todisco [4], Marques *et al.* [5], Smout and Gorantiwar [6], Gorantiwar and Smout [1]).

This paper presents two decision models which aim to minimize diffuse pollution, which is represented here as NO_3-. These models were developed under the research project POCTI/AGR/57719/2004 [3] and use the information gathered on the soil and aquifer characterization during the course of the project, as well as the results of field experiments about irrigation and land use practices.

2 The case study

The case study area is located in the watershed included in "Infrastructure 12" of the irrigated area of Alqueva Dam (Figure 1). The irrigated plots (P1, P2, P3, P4 and P5) are depicted in the same figure.

A numerical model for simulating groundwater flow has been developed, and it allowed the identification of flow direction to the small rivers (southeast to northeast) as illustrated by Figure 2 and detailed in Figure 3 and Figure 4, which show the variation of the piezometric levels between 80 m and 100 m.

The use of (http://www.modflow.com/modpath/modpath.html) MODPATH software made it possible to visualize particle movement, as displayed in Figure 5, which confirms the previous conclusions in terms of flow directions.

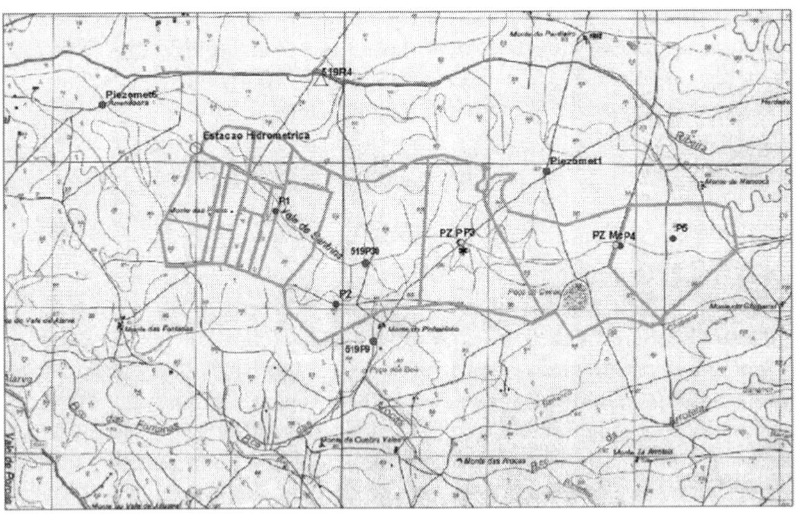

Figure 1: Case study area.

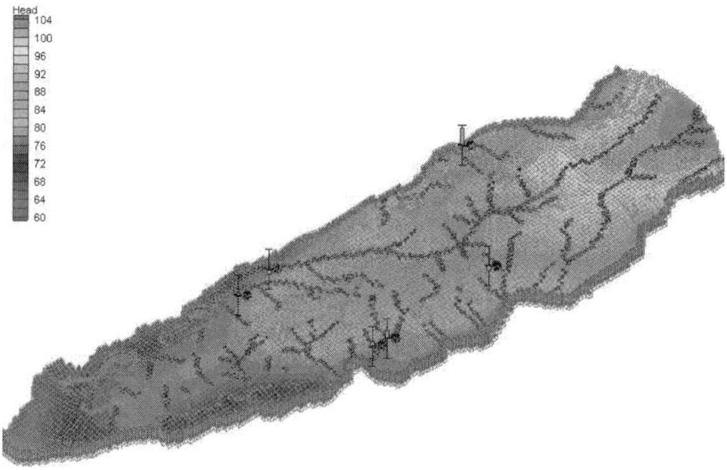

Figure 2: Piezometric levels.

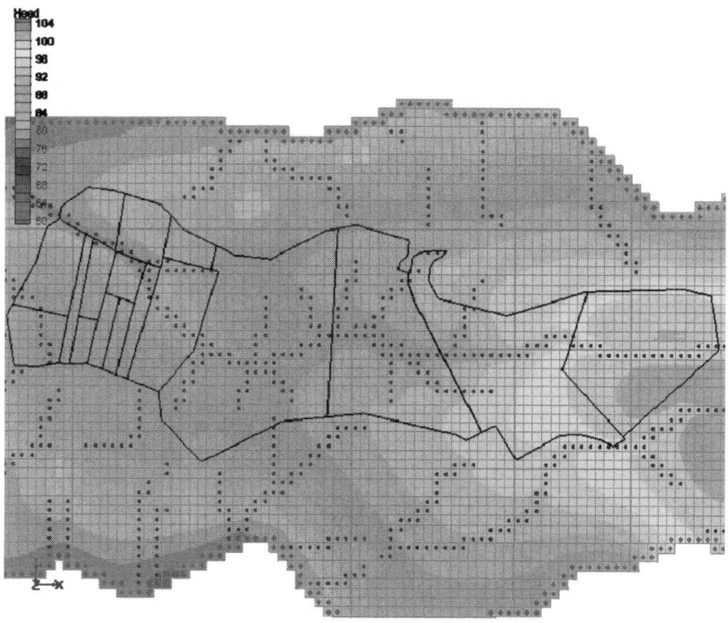

Figure 3: Detailed representation of the piezometric levels.

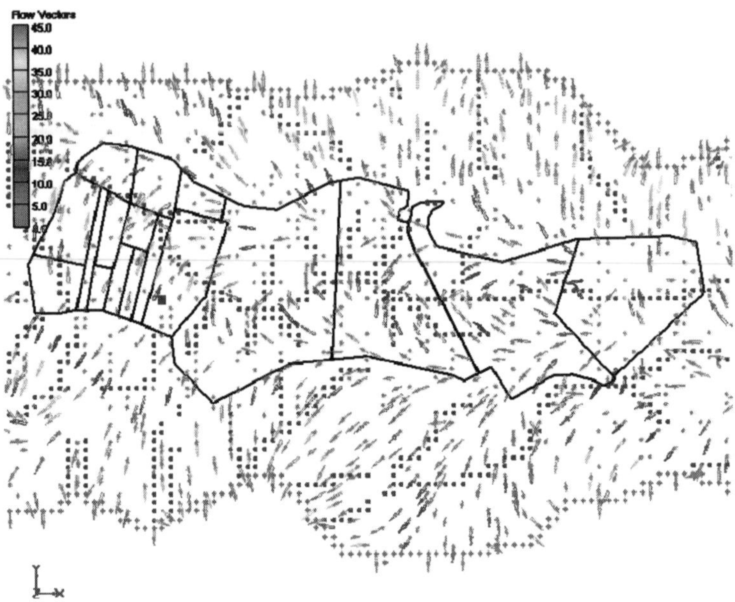

Figure 4: Intensity and direction of groundwater flow.

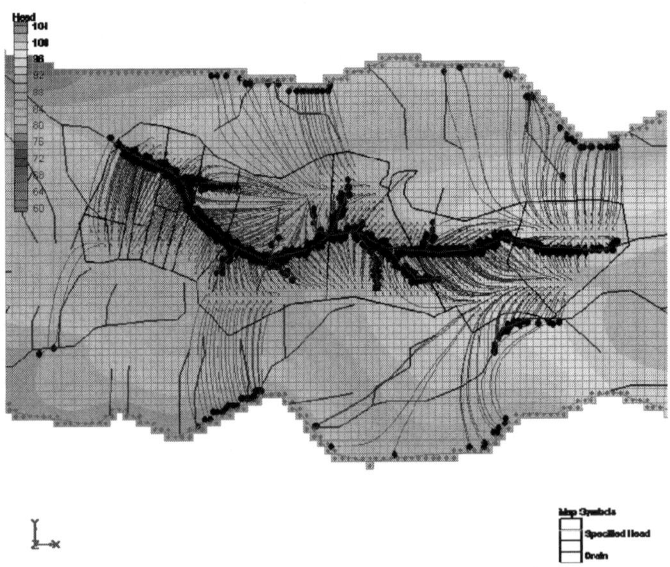

Figure 5: Groundwater particle movements.

The water quality simulation used the MT3D model that runs in the GMS interface (http://www.modflow.com/mt3d/mt3d.html).

This model simulates mass transport in complex hydrogeological environments. This model was used to evaluate the effects (in terms of nitrates) on the vadose zone and aquifers of irrigating a variety of crops which used different fertilizers applied according to different regimens.

3 Decision models and results

The decision models developed allow the information derived from applying the above-described models to be used coherently and systematically. They take into account the procedures and information produced by the simulation models mentioned, after applying them to the farmland under study. Simulations were performed to show the behaviour of the farmland for the period 2007/2015 in terms of nitrogen use/uptake by the soil/final NO3- value resulting from possible scenarios of crop rotation implemented in each plot. Two decision models were constructed: the first applies to decisions that only take into account the quality indicators in the latest growing period (2015); the second model applies to decisions relating to crops in each plot, changing from period to period. The quality indicator chosen for the decision process relates to the amount of residual NO3-in the soil throughout and at the end of the vegetative life cycle.

3.1 Decision model 1

The formulation of the first decision model is set out below. It makes it possible to establish which of various simulated crops (maize, olive trees, honeydew melons, tomatoes and cantaloupe melons) is best suited to each of the plots of farmland. The objective function of the model is given by:

$$Min\sum_i\sum_j a_{ij}y_{ij}$$

in which a_{ij} represents the response in terms of NO_3- from plot i to the growing of crop j and y_{ij} is a binary variable which takes the value one if crop j is grown in plot i and zero otherwise.

This objective is subject to a set of constraints. The first constraint represents the principle that only one crop can be grown on each plot:

$$\sum_i y_{ij} = 1, \forall j$$

The second constraint expresses the principle that all the available crops should be grown:

$$\sum_j y_{ij} = 1, \forall i$$

The next constraint expresses the limit, in average terms, which can be imposed on the cultivation as a result of applying fertilizer:

$$\frac{\sum_i \sum_j a_{ij} y_{ij}}{NP} \le NO3med_{\max}$$

in which NP is the number of plots and $NO3med_{max}$ is the maximum amount of NO_3- allowed on the farmland.

The variables y_{ij} can only take the value zero or one.

$$y_{ij} \in \{0, 1\}$$

The application of this model to the farmland being studied indicates the following crop distribution to minimise the residual amount of NO_3-:

Plot	Crop
P1	Maize
P2	Olives
P3	Honeydew melons
P4	Tomatoes
P5	Cantaloupe melons

3.2 Decision model 2

The use of the second decision model, whose formulation is set out below, makes it possible to establish which of various simulated crops is best suited to each of the plots of farmland throughout the operating period of the farm. The objective function of the model is given by:

$$Min \sum_i \sum_j \sum_k a_{ijk} y_{ijk}$$

in which a_{ijk} represents the response in terms of NO_3-from plot i to growing crop j in period k and y_{ijk} is a binary variable that takes the value one if crop j is grown in plot i in period k and value zero otherwise.

This objective is subject to a set of constraints. The first constraint represents the principle that only one crop, j, can be grown on each plot i in each period k:

$$\sum_i y_{ijk} = 1, \forall j, k$$

The second constraint expresses the principle that all the available crops (maize, olives, honeydew melons, tomatoes and cantaloupe melons) must be grown, each in its own plot) in any period, k.

$$\sum_j y_{ijk} = 1, \forall i, k$$

The next constraint expresses the limit, in average terms, which can be imposed on the cultivation as a result of applying fertilizer:

$$\frac{\sum_i \sum_j \sum_k a_{ijk} y_{ijk}}{NP * NT} \leq NO3med_{\max}$$

in which *NP* is the number of plots and *NT* is the number of growing periods.

In addition, this model allows a constraint to be considered regarding the limit that can be imposed on cultivation as a result of applying fertilizer, taking each plot individually, in each growing period:

$$\frac{\sum_j \sum_k a_{ijk} y_{ijk}}{NT} \leq NO3medpar_{\max}, \forall i$$

in which $NO3medpar_{max}$ represents the maximum average limit of NO_3-per plot, which can be imposed as a result of applying fertilizer; another constraint on the average value for the farmland in each growing period:

$$\frac{\sum_i \sum_j a_{ijk} y_{ijk}}{NT} \leq NO3medanual_{\max}, \forall k$$

in which $NO3medanual_{max}$ represents the maximum average limit of NO_3- in each period on the farm, which can be imposed as a result of applying fertilizer.

The variables y_{ij} can only take the value zero or one.

$$y_{ijk} \in \{0, 1\}$$

The resolution of this model, taking equal values for the limits $NO3med_{max}$, $NO3medpar_{max}$ and $NO3medanual_{max}$ makes it possible to determine the optimum distribution of crops per plot over all the growing periods, as shown in Table 1.

A detailed analysis of the problem leads to the conclusion that the limit value of *NO3medparmax* is the most restrictive. If a limit 35% higher than the others is used it is possible to get a global value for the objective function that is better than that obtained previously, with the corresponding distribution of crops, per plot and per period, being that shown in Table 2.

Table 1: Model 1 results.

Plots	2007	2008	2009	2010	2011	2012	2013	2014	2015
	Crops								
P1	Cant.Melon	Hon.Melon	Maize	Cant.Melon	Cant.Melon	Cant.Melon	Hon.Melon	Olives	Olives
P2	Hon.Melon	Maize	Hon.Melon	Hon.Melon	Maize	Olives	Cant.Melon	Cant.Melon	Tomatoes
P3	Maize	Cant.Melon	Cant.Melon	Maize	Hon.Melon	Hon.Melon	Olives	Hon.Melon	Cant.Melon
P4	Tomatoes	Tomatoes	Tomatoes	Olives	Olives	Maize	Maize	Maize	Hon.Melon
P5	Olives	Olives	Olives	Tomatoes	Tomatoes	Tomatoes	Tomatoes	Tomatoes	Maize

Table 2: Model 2 results.

Plots	2007	2008	2009	2010	2011	2012	2013	2014	2015
	Crops								
P1	Cant.Melon	Hon.Melon	Maize	Cant.Melon	Cant.Melon	Cant.Melon	Cant.Melon	Cant.Melon	Cant.Melon
P2	Maize	Maize	Hon.Melon	Maize	Maize	Maize	Hon.Melon	Hon.Melon	Hon.Melon
P3	Hon.Melon	Cant.Melon	Cant.Melon	Hon.Melon	Hon.Melon	Hon.Melon	Maize	Maize	Maize
P4	Tomatoes	Tomatoes	Tomatoes	Olives	Olives	Olives	Olives	Olives	Olives
P5	Olives	Olives	Olives	Tomatoes	Tomatoes	Tomatoes	Tomatoes	Tomatoes	Tomatoes

4 Conclusions

Sustainable land use and agricultural production practices are challenging when it comes to implementing the Water Framework Directive. In fact, diffuse pollution caused by agricultural practices is a major environmental problem. Therefore, it is very important to define technically sound measures to achieve good practices. Decision models can play an important role here. The decision models presented are very promising since they allow the definition of land use and farming practices that can minimize the effects of diffuse pollution in aquifers in order to comply with the limits set by the applicable legislation.

Acknowledgement

The authors gratefully acknowledge support received from the Fundação para a Ciência e a Tecnologia through grant POCTI/AGR/57719/2004.

References

[1] Gorantiwar, S.D, Smout, I.K. (2005), "Multilevel approach for optimizing land and water resources and irrigation deliveries for tertiary units in large irrigation schemes. I: Application", *J. Irrig. Drain. Eng.*, 131(3), pp. 264–272.

[2] Kumar, D.N., Raju, K.S., Ashok, B. (2006), "Optimal Reservoir operation for irrigation of multiple crops using genetic algorithms", *J. Irrig. Drain. Eng.*, 132(2), pp. 123–129.

[3] Leitão, T.E., Laranjeira, I., Oliveira, L., Paralta E., Lobo Ferreira, J.P., Cunha, M.C. and Terceiro, P. (2009), "Metodologias para um melhor planeamento e gestão do uso do solo agrícola atendendo à vulnerabilidade dos aquíferos à poluição difusa", *Relatório Final. Relatório 301/2009-NAS*, 206 pp.

[4] Mannocchi, F., Todisco, F. (2006), "Optimal reservoir operations for irrigation using a three spatial scales approach", *J. Irrig. Drain. Eng.*, 132(2), pp. 130–206.

[5] Marques, F.M., Lund, J.R., Howitt, R.E. (2005), "Modelling irrigated agricultural production and water use decisions under water supply uncertainty", *Water Resour. Res.*, 41(3), W08423.

[6] Smout, I.K., Gorantiwar, S.D. (2005), "Multilevel approach for optimizing land and water resources and irrigation deliveries for tertiary units in large irrigation schemes. I: Method", *J. Irrig. Drain. Eng.*, 131(3), pp. 254–263.

Correlation between chemo-physical parameters NH_3, NO_2, NO_3, SO_4, pH and aquifer and impervious layer electrical resistance in groundwater

G. Badalians Gholikandi[1] & H. R. Orumieh[2]
[1]*Power and Water University of Technology (PWUT), Water Research Institute (WRI), Iran*
[2]*Pars arianab Consulting Engineers, Iran*

Abstract

Groundwater resources polluted by different compounds limits the possibility of using this important resource. In this research, with the intention of improving the controls for water quality, geo electric methods and determination of the relation between the chemo-physical properties in groundwater and the aquifer and impervious layer of electrical resistances were investigated. The wells in the Yaftabad area in Tehran were investigated in this research. The results show that the true resistance of the aquifer is about 90–134 ohm.m, and for the impervious layer it is 19–78 ohm.m. According to the aquifer and impervious layer resistance and the results of laboratory study diagrams, equations and correlations between the chemo-physical properties of groundwater and true specific electrical resistance were obtained. There is a relation between the true specific electrical resistance of the aquifer and the impervious layer with the most numerous of chemo-physical properties. The extent of the correlation constant variation for NO_2, NO_3, NH_3, and pH is medium and for SO_4 is strong with true specific electrical resistance of the aquifer. The extent of the correlation constant variation for NO_2 is medium, for NH_3 and SO_4 is strong and for NO_3 and pH is too strong with true specific electrical resistance of the impervious layer. The average extent value of the correlation constant variation between chemo-physical properties in groundwater and the true specific electrical resistance of the aquifer is equal to 0.698 and for the impervious layer is equal to 0.539.

Keywords: groundwater quality, aquifer, impervious layer, geo electrics.

WIT Transactions on Ecology and the Environment, Vol 134, © 2010 WIT Press
www.witpress.com, ISSN 1743-3541 (on-line)
doi:10.2495/SI100091

1 Introduction

Groundwater is a highly useful and often abundant resource. It is also ecologically important. About half the population in the city of Tehran relies to some extent on groundwater as a source of drinking water, and still more use it to supply their factories with process water or their farms with irrigation water. Certain problems have beset the use of groundwater around the city and the province of Tehran.

Some of the most important parameters in water are nitrogen compounds. It contains Nitrite (NO_2^-), Nitrate (NO_3^-) and ammonium (NH_3). Nitrate is the common form of inorganic nitrogen found dissolved in water. Groundwater in this region can be polluted by nitrogen from both the discharge (diffusion) of municipal wastewater and drainage from agricultural lands. The health hazard of ingesting excessive nitrate in water is infant methemoglobinemia [3].

Since sulfate is soluble in water, it is found at high concentrations in many aquifers. Sulfate is an oxidized form of sulfur. There are many potential sources of sulfate. Gypsum is an important source in many aquifers that have high concentrations of sulfate. Reduced forms of sulfur are oxidized to sulfate in the presence of oxygen. This process often occurs when sulfide minerals are mined.

Sulfate does not have a health-based drinking water standard. Sulfate has laxative effects and imparts an unpleasant taste to water. Aquifers with high concentrations of hydrogen sulfide have a bad odor [6].

The pH of water determines the solubility and biological availability of chemical constituents such as nutrients and heavy metals [10].

Total dissolved solids (TDS), is defined as the concentration of all dissolved minerals in the water. TDS are a direct measurement of the interaction between ground water and subsurface minerals. High TDS, greater than 1000 mg/L, is commonly objectionable or offensive to taste. The high TDS may cause corrosion of pipes and plumbing systems [8].

Ground water, especially if the water is acidic, in many places contains excessive amounts of iron. Iron causes reddish stains on plumbing fixtures [9]. Table 1 shows total maximum acceptable concentration of chemo-physical properties based on world standards.

In this research, regarding the importance and danger of groundwater pollutants in urban zones and the need to get improved control of wells quality, we investigated geoelectric methods and determination of some relation between chemo-physical properties in groundwater and aquifer and impervious layer electrical resistances.

Geoelectrical methods are used extensively in groundwater mapping for investigation of the vulnerability of aquifers and shallow aquifers themselves. A geoelectrical measurement is carried out by recording the electrical potential arising from current input into the ground with the purpose of achieving information on the resistivity structure in the ground.

Geoelectrical data are commonly expressed as apparent resistivities

$$\rho a = (\Delta V/l)K$$

Table 1: Total maximum acceptable concentration of chemo-physical parameters, based on world famous standards.

Parameter	USEPA Maximum Contaminant Level(MCL)	Canada Maximum Acceptable Concentration	EEC Maximum Admissible Concentration	Japan Maximum Admissible concentration	WHO Guideline
Ammonium			0.5 mg/L	No standard	1.5 mg/L
Cadmium	0.005 mg/L	0.005 mg/L	0.005 mg/L	0.01 mg/L	0.01 mg/L
Chromium	0.1 mg/L	0.05 mg/L	0.05 mg/L	0.05 mg/L	0.05 mg/L
Iron	0.3 mg/L	0.3 mg/L	0.2 mg/L	0.3 mg/L	0.3 mg/L
Manganese	0.05 mg/L	0.05 mg/L	0.05 mg/L	0.01-0.05 mg/L	0.1-0.5 mg/L
Nitrate/Nitrite,total	10 mg/L as N			10 mg/L as N	
Nitrates	10 mg/L as N	10 mg/L as N	50 mg/L	10 mg/L as N	50 mg/L as NO3-
Nitrites	1 mg/L as N	3.2 mg/L	0.1 mg/L	10 mg/L	3 mg/L as NO2-
pH	6.5-8.5	6.5-8.5	6.5-9.5	5.8-8.6	6.5-8.5
Potassium		0.002 mg/L	0.5 µg/L C6H5OH	0.005 mg/L	
Solids,total dissolved	500 mg/L	500 mg/L	No Standard	500 mg/L	1000 mg/L
Sodium			75-150 mg/L	200 mg/L	200 mg/L
Sulfate	250 mg/L	500 mg/L	250 mg/L	No standard	250 mg/L
Zinc	5 mg/L	5 mg/L	No Standard	1.0 mg/L	3.0 mg/L

where ΔV is the measured potential, l the transmitted current and K the geometrical factor. One of the most common electrode arrays is Schlumberger.

Vertical electrical sounding (VES) is used to determine the resistivity variation with depth. A VES is typically carried out in Schlumberger array, where the potential electrodes are placed in a fixed position with a short separation and the current electrodes are placed symmetrically on the outer sides of the potential electrodes (Fig. 1). After each resistivity measurement the current electrodes are moved further away from the center of the array. In this way the current is stepwise made to flow through deeper and deeper parts of the ground. The positions of the current electrodes are typically logarithmically distributed with at least 10 positions per decade [4, 5, 7].

The most important parameter investigated in this paper is Correlation constant. Correlation constant is a statistical parameter that shows different degrees and conditions of correlation between 2 variations. General formula for calculating correlation constant between 2 variations X and Y is like below. This constant (R_{XY}) is named Pearson constant [4, 5, 7].

$$R_{XY} = \frac{\Sigma(X - M_x)(Y - M_Y)}{\sqrt{\Sigma(X - M_X)^2 \Sigma(Y - M_Y)^2}}$$

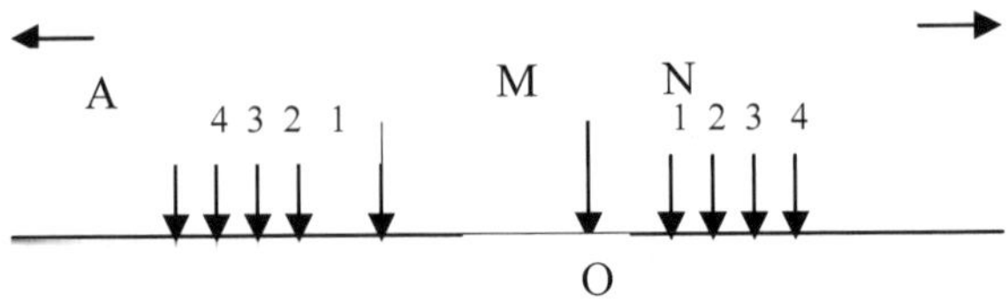

Figure 1: Flow electrodes movements' stages in electrical sounding method (Schlumberger array).

Table 2: Different values of correlation constants and their qualifying classifications.

Constant Value lower than 0.2	Insignificant
Constant Value between 0.2 to 0.4	Weak
Constant Value between 0.4 to 0.6	Medium
Constant Value between 0.6 to 0.8	Strong
Constant Value between 0.8 to 1	Too Strong

If validity of existing statistical relation assured between parameters, qualify assessment will define intensity of this relation. Correlation constants can classify as following (Table 2).

2 Materials and methods

In order to analyze the water quality, laboratory experiments have been carried out using the APHA 1998 [1].

Wells in Yaftabad area in Tehran (Group wells no 300, which include wells no: 301, 302, 303, 304, 305 and 306) were investigated as a case study (Fig. 2).

For studying electrical properties of aquifer, impervious layer and their resistance, we completed six electrical soundings (in Schlumberger method) around 301 to 306 application wells with an average length of AB equal 600 meter. This act accomplished three times for every well except 302, and for well no 302, because of better circumference five times at different time periods [2].

3 Results

Sounding measurements results comment on comparison with standard diagrams. The result of these comments is showing that the true resistance of the aquifer is about 90 to 134 ohm.m and for the impervious layer are 19 to 78 ohm.m. According to aquifer and impervious layer resistance, also laboratory studies we obtained following results (Figures 3, 4).

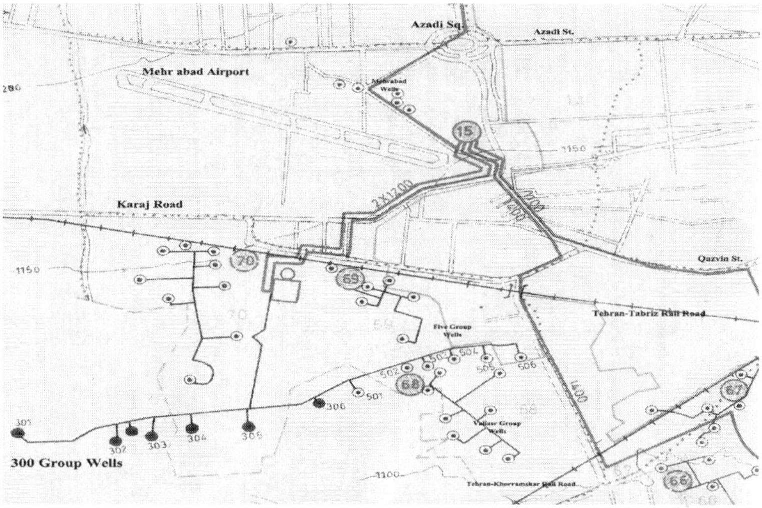

Figure 2: Settlement of 300 group wells [2].

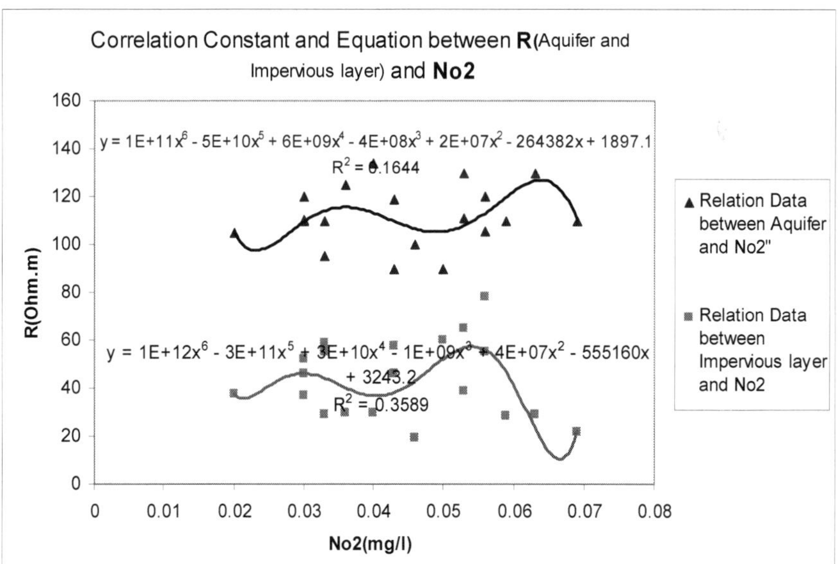

Figure 3: Diagrams, equations and correlations.

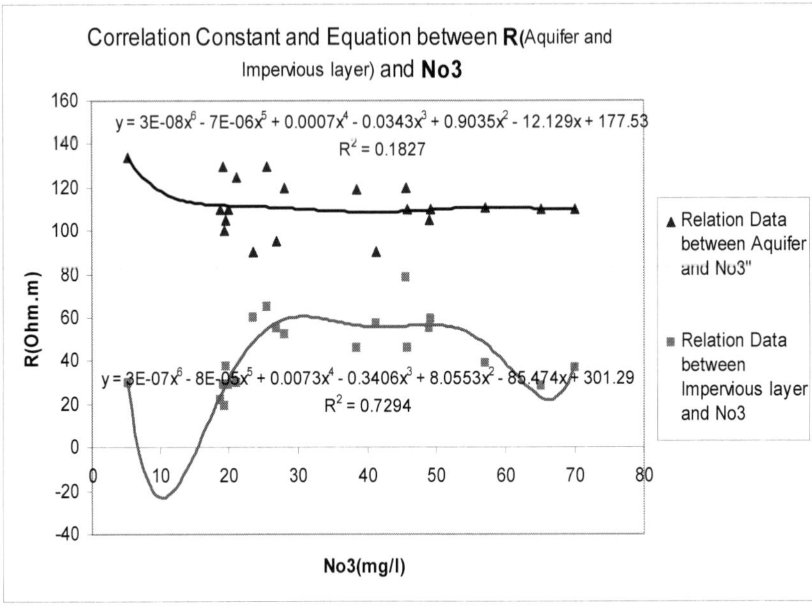

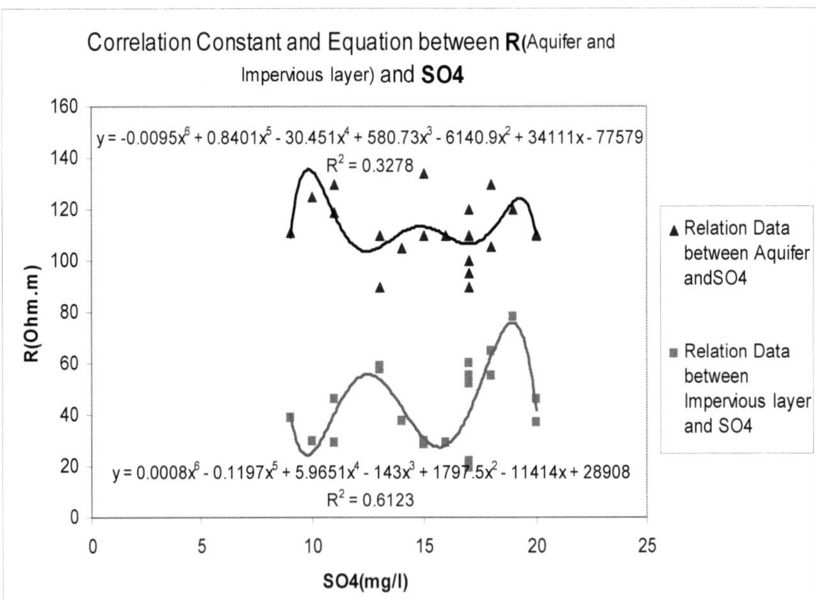

Figure 3: Continued.

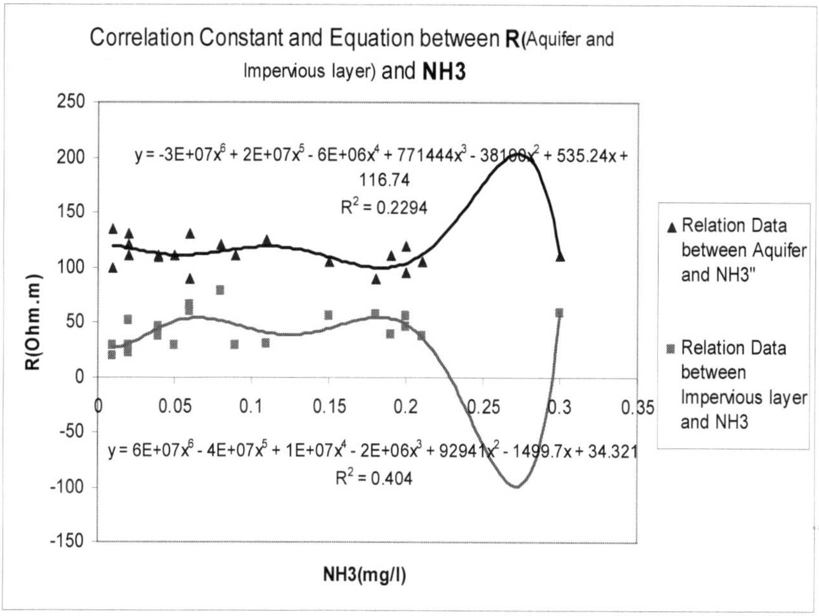

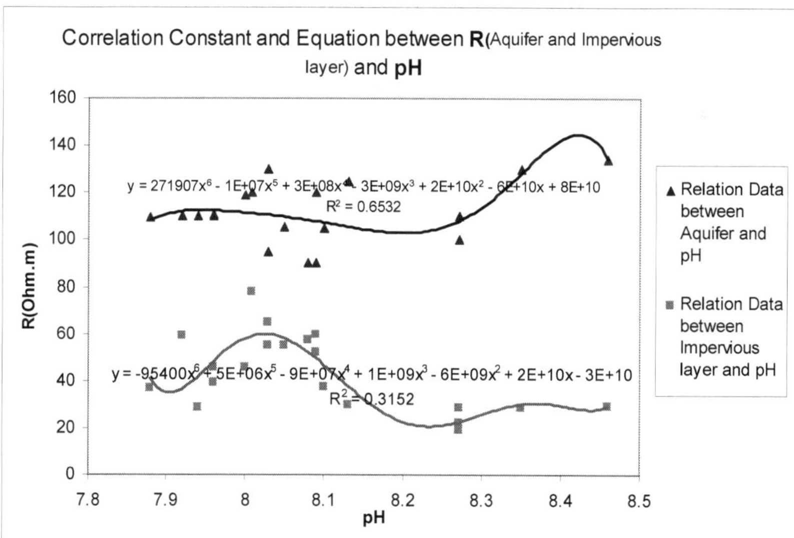

Figure 3: Continued.

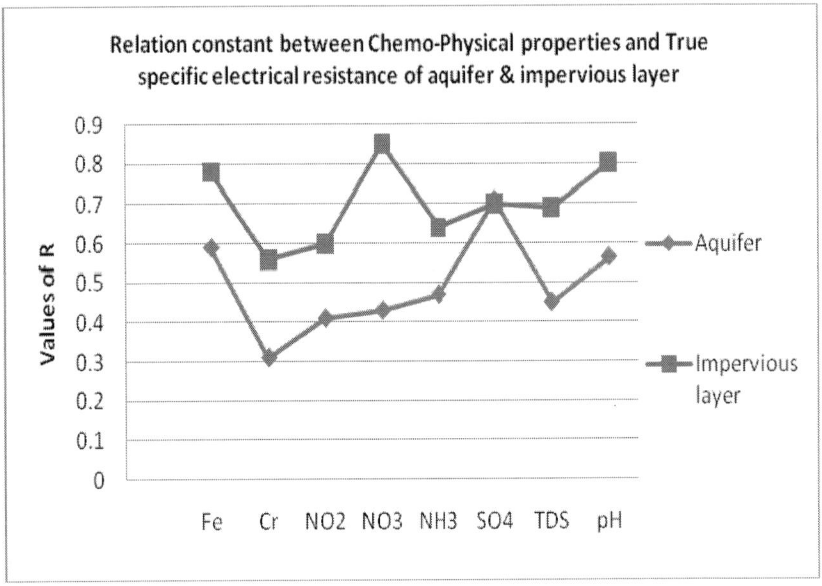

Figure 4: Calculated correlation values between true specific electrical resistance of the aquifer and the impervious layer with the chemo-physical properties of groundwater.

4 Conclusion

After this study and regarding to the results, there is a relation between true specific electrical resistance of aquifer and impervious layer with the most numerous of chemo-physical properties. Except Cr extent of correlation constant variation for Fe, NO_2, NO_3, NH_3, TDS and pH are medium and for SO_4 is strong with true specific electrical resistance of aquifer. Except K, extent of correlation constant variation for NO_2 and Cr are medium and for NH_3, SO_4 and TDS are strong; NO_3 and pH are too strong with true specific electrical resistance of impervious layer. Average extent values of correlation constant variation between chemo-physical properties in groundwater & true specific electrical resistance of aquifer is equal 0.698 and for impervious layer is equal 0.539.

Acknowledgements

This study was supported by Tehran Water and Wastewater Company (TWWC), National Water and Wastewater Company Iran (NWWC) and Water Research Institute (WRI).

References

[1] American Public Health Association (APHA), (1998). Standard Method for the Examination of Water and Wastewater, 20th ed., Washington DC, USA.

[2] Badalians Gholikandi, G., Kahpood, H.R. (2005). Feasibility study of geoelectrical methods application for Groundwater quality monitoring, Power and Water University of Technology, Tehran, Iran.
[3] Badalians Gholikandi, G. (2003). Water chemistry, Nowpardazan Publishing company, Tehran, Iran.
[4] Christensen, NB., Sorensen, KI. (1998). Surface and borehole electric and electromagnetic methods for hydrogeological investigations, European Journal of Environmental and Engineering Geophysics 3, 75-90.
[5] Ernstson, K., Kirsch, R. (2006). Geoelectrical methods, basic principles. In Kirsch R (ed.): Groundwater Geophysics, A Tool for Hydrology, 85-108, Springer.
[6] Minnesota Pollution Control Agency, (1999). Sulfate in Minnesota's Groundwater, Saint Paul, USA.
[7] Moller, I., Sorensen, KI., Auken, E. (2009). Geoelectrical methods, www.burval.org/final/pdf/077-088.pdf.
[8] Ohiodnr (ODNR), (1987). Groundwater quality, Division of water, FS 97-47, Columbus, Ohio, USA.
[9] US Geological Survey (USGS), (2009). Quality of groundwater, Office of Groundwater, http://pubs.usgs.gov./gip/gw/quality.html.
[10] Water on the Web (WOW), (2008). Water quality parameters, pH, http://www.waterontheweb.org/under/waterquality/pH.html.

Theoretical basis and significance of variance of discharge as a bidimensional variable in microirrigation lateral design

E. R. Detomini[1,2] & J. A. Frizzone[3]
[1]*Ministry of Planning, Budget and Management (Secretariat for National Assets), Brazil*
[2]*University of Cuiabá, Brazil*
[3]*University of São Paulo, Brazil*

Abstract

In order to support the theoretical basis and contribute to the improvement of educational capability issues relating to irrigation systems design, this work presents an alternative deduction for variance of the discharge as a bidimensional and independent random variable. Then, a subsequent brief application of an existing model is applied for statistical design of laterals in micro-irrigation. The better manufacturing precision of emitters allows lengthening of a lateral for a given soil slope, although this does not necessarily mean that the statistical uniformity throughout the lateral will be more homogenous.

Keywords: error theory, statistical approach, manufacturing coefficient.

1 Introduction

In microirrigation systems, generally the longer the lateral pipeline implemented in a sub-main unit is, the less the final cost of the designed project and hence the overall profitability of the system is improved. In practice, two approaches may be used to optimise the lateral length while maintaining the desirable uniformity of applied water: hydraulic and statistical. Considering the latter, the coefficient of variation of pressure head (*CVHp*) term appears in the model due to the concept of discharge variance, where the source-point discharge is assumed to be susceptible to oscillations in both the pressure head (*H*) and emitter coefficient (*K*). The latter is inherent to the existing variability of emitters coming from the

WIT Transactions on Ecology and the Environment, Vol 134, © 2010 WIT Press
www.witpress.com, ISSN 1743-3541 (on-line)
doi:10.2495/SI100101

manufacturing processes. Specifically, discharge variation values are assumed to follow a bivariate normal distribution without correlation between independent variables and hence these oscillations need to be contemplated when designing laterals. Small differences between what appear to be identical emitters may result in significant discharge variations.

The manufacturer's coefficient of emitter variation is a measure of the variability of discharge of a random sample of a given make, model and size of emitter, as produced by the manufacturer and before any field operation or ageing has taken place (ASAE [2]). It reflects the degree of precision with which the emitters are made by the manufacturer. With respect to average source-point discharge, one can use Taylor's series for deducting the variance with some difficulty degree or, alternatively, it may be deduced more easily via error theory. In order to support the theoretical basis and contribute to the improvement of educational capability issues relating to irrigation design, this work uses a simple engineering algebraic manipulation and thereby presents an alternative deduction of the variance of discharge as a bidimensional variable and a brief application of Anyoji and Wu's [1] model for statistically design laterals in micro-irrigation.

2 Model description and alternative deduction

To describe the emitter discharge (q, L h^{-1}), Keller and Bliesner [8] suggest the widely used power equation as a function of pressure head (H, m):

$$q = K \cdot H^{x} \tag{1}$$

The variance of discharge (δ_q^2) throughout lateral is given by:

$$\delta_q^2 = \frac{1}{N} \cdot \sum_{i=1}^{N} (q_i - q_m)^2 \quad ; \quad \delta_q^2 = \frac{1}{N} \cdot \sum_{i=1}^{N} (dq)^2 \tag{2}$$

in which q_i is the discharge of the ith emitter; q_m is the mean discharge; and dq is the deviation of discharge values (continuous variable); and N is the number of emitters along a lateral.

Likewise, the variances of independent variables are:

$$\delta_K^2 = \frac{1}{N} \cdot \sum_{i=1}^{N} (dK)^2 \tag{3}$$

$$\delta_H^2 = \frac{1}{N} \cdot \sum_{i=1}^{N} (dH)^2 \tag{4}$$

in which dq, dK and dH are the deviations of discharge, emitter constant and pressure head, respectively.

The coefficient of correlation (ρ) between the random variables K and H measures the magnitude and direction in which they linearly straight ahead together (Moore [10]), being possible to present it as:

$$\rho = \frac{\sum_{i=1}^{N}(K_i - K_m)\cdot(H_i - H_m)}{\sqrt{\sum_{i=1}^{N}(K_i - K_m)^2}\cdot\sqrt{\sum_{i=1}^{N}(H_i - H_m)^2}} \tag{5}$$

Rearranging eq. (5) in a convenient manner:

$$\sum_{i=1}^{N}(dK)\cdot(dH) = \rho\cdot\sqrt{\sum_{i=1}^{N}(K_i - K_m)^2}\cdot\sqrt{\sum_{i=1}^{N}(H_i - H_m)^2} \tag{6}$$

According to the error theory, the total derivative of eq. (1) corresponds to the total error implied when it is used. Particularly for q, this may be described by:

$$dq = \frac{\partial q}{\partial K}\cdot dK + \frac{\partial q}{\partial H}\cdot dH \tag{7}$$

where:

$$\frac{\partial q}{\partial H} = K_m \cdot x \cdot H_m^{x-1} \tag{8}$$

$$\frac{\partial q}{\partial K} = H_m^x \tag{9}$$

Whereas the higher-order partial differentiations are given by:

$$\frac{\partial^2 q}{\partial K^2} = 0 \tag{10}$$

$$\frac{\partial^2 q}{\partial H^2} = K_m \cdot x \cdot (x-1)\cdot H_m^{x-2} \tag{11}$$

Substituting eq. (7) in eq. (2):

$$\delta_q^2 = \frac{1}{N}\cdot\sum_{i=1}^{N}\left(\frac{\partial q}{\partial K}\cdot dK + \frac{\partial q}{\partial H}\cdot dH\right)^2 \tag{12}$$

Expanding eq. (12) by taking the binomial quadratic expansion:

$$\delta_q^2 = \frac{1}{N}\cdot\sum_{i=1}^{N}\left[\left(\frac{\partial q}{\partial K}\cdot dK\right)^2 + 2\cdot\frac{\partial q}{\partial K}\cdot dK\cdot\frac{\partial q}{\partial H}\cdot dH + \left(\frac{\partial q}{\partial H}\cdot dH\right)^2\right] \tag{13}$$

$$\delta_q^2 = \frac{\sum_{i=1}^{N}\left(\frac{\partial q}{\partial K}\cdot dK\right)^2}{N} + \frac{\sum_{i=1}^{N}2\cdot\frac{\partial q}{\partial K}\cdot\frac{\partial q}{\partial H}\cdot dK\cdot dH}{N} + \frac{\sum_{i=1}^{N}\left(\frac{\partial q}{\partial H}\cdot dH\right)^2}{N} \tag{14}$$

All terms involving partial differentiation on the equation above are possible to be taken out of sum operators because they do not contemplate indices i[th]. This might be verified by reviewing eq. (8) and eq. (9). Thus:

$$\delta_q^2 = \left(\frac{\partial q}{\partial K}\right)^2 \cdot \frac{\sum_{i=1}^{N} (dK)^2}{N} + 2 \cdot \frac{\partial q}{\partial K} \cdot \frac{\partial q}{\partial H} \cdot \frac{\sum_{i=1}^{N} dK \cdot dH}{N} + \left(\frac{\partial q}{\partial H}\right)^2 \cdot \frac{\sum_{i=1}^{N} (dH)^2}{N} \tag{15}$$

Substituting eq. (6) on eq. (15):

$$\delta_q^2 = \left(\frac{\partial q}{\partial K}\right)^2 \cdot \frac{\sum_{i=1}^{N} (dK)^2}{N} + 2 \cdot \frac{\partial q}{\partial K} \cdot \frac{\partial q}{\partial H} \cdot \frac{\rho \cdot \sqrt{\sum_{i=1}^{n} (K_i - K_m)^2} \cdot \sqrt{\sum_{i=1}^{n} (H_i - H_m)^2}}{N} + \left(\frac{\partial q}{\partial H}\right)^2 \cdot \frac{\sum_{i=1}^{N} (dH)^2}{N} \tag{16}$$

Assuming that the variables K and H are independent each other, especially when system pressure is low and emitter material is rigid enough to not provoke deformation, ρ is zero and the second term of second member vanishes in the equation above. Hence, eq. (16) reduces to:

$$\delta_q^2 = \left(\frac{\partial q}{\partial K}\right)^2 \cdot \frac{\sum_{i=1}^{N} (dK)^2}{N} + \left(\frac{\partial q}{\partial H}\right)^2 \cdot \frac{\sum_{i=1}^{N} (dH)^2}{N} \tag{17}$$

Substituting equations (3), (4), (8) and (9) on the equation above, we obtain:

$$\delta_q^2 = \left(H_m^x\right)^2 \cdot \delta_K^2 + \left(K_m \cdot x \cdot H_m^{x-1}\right)^2 \cdot \delta_H^2 \tag{18}$$

Providing that the coefficients of variation are considered, it highlights that:

$$CVq = \frac{\sqrt{\delta_q^{\,2}}}{q_m} \tag{19}$$

$$CV_K = \frac{\sqrt{\delta_K^{\,2}}}{K_m}; \quad \text{or} \quad \delta_K^{\,2} = \left(K_m \cdot CV_K\right)^2 \tag{20}$$

$$CVH = \frac{\sqrt{\delta_H^{\,2}}}{H_m}; \text{ or } \delta_H^{\,2} = \left(H_m \cdot CVH\right)^2; \text{ or } CVH^2 = \frac{\delta_H^{\,2}}{H_m^2} \tag{21}$$

Substituting eq. (20) and eq. (21) on eq. (18):

$$\delta_q^2 = \left(H_m^x\right)^2 \cdot \left(K_m \cdot CV_K\right)^2 + \left(K_m \cdot x \cdot H_m^{x-1}\right)^2 \cdot \left(H_m \cdot CVH\right)^2 \tag{22}$$

Reorganising the equation above in a convenient way:

$$\delta_q^{\,2} = K_m^2 \cdot H_m^{2 \cdot x} \cdot \left(CV_K^2 + x^2 \cdot CV_H^2\right) \tag{23}$$

The deducted equation above is going to be utilised in the coming steps, and represents the variance of point-source of the discharge as a bidimensional variable, that depends upon both emitter coefficient (K) and pressure head (H) if clogging and temperature effects on emitter discharge are neglected (i.e. design stage).

3 Applicability of variance of the bidimensional discharge

The expectation $[E(q)]$ of q utilising a Taylor's series expansion is:

$$E(q) = K_m \cdot H_m^x + \frac{1}{2} \cdot \left[\frac{\partial^2 q}{\partial K^2} \cdot \delta_K^2 + \frac{\partial^2 q}{\partial H^2} \cdot \delta_H^2 \right] \tag{24}$$

When all events have the same probability of occurrence, the expectation (or expected value) equals to arithmetic average of the values, but the second term of second member stand for the variation of discharge due to the effect of both K and H variation. The expectation of q can also be called q_m.

Substituting equations (10), (11) and (21), into (24) becomes:

$$q_m = K_m \cdot H_m^x \cdot \left[1 + 0.5 \cdot x \cdot CVH^2 \cdot (x-1) \right] \tag{25}$$

Substituting both eq. (25) and eq. (23) in eq. (19), the latter comes to:

$$CVq = \frac{\sqrt{K_m^2 \cdot H_m^{2 \cdot x} \cdot \left(CV_K{}^2 + x^2 \cdot CVH^2 \right)}}{K_m \cdot H_m^x \cdot \left[1 + 0.5 \cdot x \cdot CVH^2 \cdot (x-1) \right]} \tag{26}$$

As done in this paper for the expectation of q (see eq. (24)), Anyoji and Wu [1] have demonstrated eq. (26) by using the approach of Lindley [9] about Taylor's Series to estimate the variance of a variable that is function of two independent variables, therefore, differently from the herein presented approach, based on error theory and fully deducted for the variance. The approach of Lindley [9] is certainly and fully valid but is more complex for deducting the variance than error theory, whereas the closed equation for deducting expectation through Taylor's Series can be achieved more easily.

The variable *CVK* is often referred as *CVf*. Additionally, *CVH* might be called as *CVHp* (coefficient of variation of pressure on project) at the design stage, most likely to be different from the *CVH* of the working irrigation system. By considering this and conveniently rearranging the equation above, we finally have the expression that, indeed, is the first step to statistically design laterals in irrigation systems:

$$CVq \cdot \left[1 + \frac{1}{2} \cdot x \cdot CVHp^2 \cdot (x-1) \right] - \sqrt{CVf^2 + x^2 \cdot CVHp^2} = 0 \tag{27}$$

Since *CVq* is a user-defined value according to a uniformity criteria, generally not more than 0.07 (or 7%), it means, an emitter flow variation of 20% (Wu [14]). The attributes *CVf* and x are known elements from manufacturers, whereas *CVHp* becomes the variable of interest. As it is not possible to explicit derive *CVHp*, it must be found through any iterative method (i.e. Newton-Raphson, Bisection or Secant).

Even though Frizzone et al [6] have utilised eq. (26) for their calculations, which is basically the same of that one developed by Anyoji and Wu [1] and the rearranged eq. (27), the latter is mathematically more suitable than the former in terms of programming because expresses a typical algebraic equation to be solved in zero [i.e. f($CVHp$) = 0]. Burden and Faiures [4] provide more details about the advantages and limitations of some of the main iterative methods in terms of facility and speed of convergence. As far as we have found, such value might be easily obtained through root-finding algorithm existing in Microsoft Excel® (Tools > Goal Seek…) with no difficulty of convergence.

If the second term inside the brackets of the denominator of eq. (26) is small to note that when x tends to 0.5 (i.e. working in full turbulence), the expression in the brackets $\left[1+0.5\cdot x\cdot CVH^2\cdot(x-1)\right]$ tends to 1 so that the expectation is provided directly by the original functional form of q with the mean values of K and H [see eq. (1)] and the CVq can be reduced to same equation of that one derived by Bralts [3].

When emitters are ideally compensating (i.e. x = 0), there is no solution for eq. (27) because this becomes meaningful. When x = 0.5 (orifice-type emitters), the value of $[1 + 0.5 \cdot x \cdot CVH^2 \cdot (x - 1)]$ tends to be higher than that one when $0 < x < 0.5$. In other words, compensating emitters are more influenced by the value of $[1 + 0.5 \cdot x \cdot CVH^2 \cdot (x - 1)]$ as a consequence of higher resulting CVH values. However, this is only noted when $[1 + 0.5 \cdot x \cdot CVH^2 \cdot (x - 1)]$ is calculated after eq. (27) is solved. When x = 1, the value of $[1 + 0.5 \cdot x \cdot CVH^2 \cdot (x - 1)]$ is zero.

After finding $CVHp$, it is used as an input value to optimise the length of laterals according to (Anyoji and Wu [1]):

$$\frac{(m+1)^2}{(2\cdot m+3)\cdot(m+2)^2}\cdot Hf^2+\frac{1}{12}\cdot\Delta Z^2+\frac{(m+1)}{(m+2)\cdot(m+3)}\cdot Hf\cdot\Delta Z-(CVHp\cdot Hm)^2=0 \tag{28}$$

in which m is the exponent with respect to discharge in friction equations; H_m is the mean value of pressure (m) throughout lateral – calculated by inverting eq. (25); Hf is the total friction loss (m) throughout lateral; and ΔZ is the level difference (m), assuming a negative value only in case of downhill slopes.

The following equations must be substituted in eq. (28) prior to solve it:

$$\Delta Z=\frac{So}{100}\cdot L \tag{29}$$

$$Hf=\frac{Khf}{3600000^m}\cdot\frac{q^m}{D^n}\cdot\left(\frac{Se+hfe}{Se^{m+1}}\right)\cdot\frac{L^{m+1}}{m+1} \tag{30}$$

in which L is the length (m) of lateral – the goal value to be found iteratively; So is the soil slope (%), assuming it as uniform; hfe the local head losses (m – equivalent length) at emitter insertions; Khf the friction constant; Se the emitter spacing (m); D the internal lateral diameter (m); n the adjusted empirical coefficient.

Analogously to eq. (27), we found no problem to solve eq. (28) by using either of most traditional methods (Newton-Raphson, Secant or bisection). In addition, it is necessary to comment that the current approach disregards higher order terms of Christiansen coefficient (F), in which $F = 1 / (m+1)$, valid for laterals contemplating a large number of emitters (i.e. N > 20). However, the exact length of lateral and consecutively the number of emitters (N) are not known *a priori* (i.e. L may be very short under high slope situations if at the same time H_m is low and CVq is rigorously low), which then would require the full equation for calculating F regardless of its former simplification.

As N must be a non-decimal number, it is worthwhile to remind that once L value is found from eq. (28), N must be truncated (for N_R) preferably towards to the nearest smaller integer value in order to obtain a little gain in terms of pressure throughout the lateral, which has the addition a benefit of preventing clogging of emitters due small particles of sediments. As a consequence, a new lateral length (L_R) is then recalculated:

$$L_R = N_R \cdot Se \tag{31}$$

Thus, the lateral inlet pressure (H_{inl}, m) and the pressure related to each i emitter (H_i, m) may be calculated, respectively, according to:

$$H_m = H_{inl} - \left(\frac{m+1}{m+2}\right) \cdot Hf_R - 0.5 \cdot \Delta Z \tag{32}$$

$$H_i = H_{inl} - \left[1 - (1 - CR_i)^{m+1}\right] \cdot Hf_R - CR_i \cdot \Delta Z \tag{33}$$

where Hf_R and ΔZ_R are the recalculated ΔZ and Hf, in an analogous way provided by equations (29) and (30), respectively, by replacing L_R instead of L. Likewise eq. (28), ΔZ is here is also assumed as negative for declivity situations; CR_i is the relative length, i.e. the length between the first and the ith emitter, is given by:

$$CR_i = \frac{i \cdot Se}{L_R} \tag{34}$$

Consequently, the discharge (q_i, L h^{-1}) in each emitter may be obtained by:

$$q_i = K \cdot H_i^x \tag{35}$$

An equation that accounts for both emitter variation and system pressure variation is one denoting the uniformity of emission, which was modified and redefined for design purposes, as pointed out by Bralts [3]. However, despite the existence of a number of equations to assess the uniformity (i.e. absolute emission uniformity, statistic uniformity, etc.), Favetta and Botrel [5] have compared all these equations and have shown a strong correlations among them, which allows the adoption of that one that most suits the user's needing. At the design stage, the statistical uniformity (US, %) of water depth can be used (Juana et al [7]), according to (Wilcox and Swailess [13]):

$$US = 100 \cdot (1 - CVq) = 100 \cdot \left(1 - \frac{\sum_{i=1}^{n}\left(q_i - \frac{1}{N} \cdot \sum_{i=1}^{N} q_i\right)}{\frac{1}{N} \cdot \sum_{i=1}^{N} q_i}\right) \tag{36}$$

There are more factors affecting the uniformity of micro-irrigation than that of sprinkler irrigation besides manufacturer's variation (Wu [14]). These other factors might be grouping of emitters, plugging and temperature. The effect of temperature on emitter flows can be neglected when a turbulent flow emitter is used (Peng et al [11]), whereas plugging of emitters is developed with respect to time and occurs in the form of partial plugging and complete plugging. As partial pluggings are difficult to evaluate, plugging evaluations are taken by using completely plugged emitters (Wu [14]).

For illustrative purposes values for length were simulated from variations in the input variable *CVf* (2%, 4%, 6%, 8% and 10%), considering a low-pressure irrigation system designed in a 2% downhill uniform slope (*So* = – 2). The emitter spacing was 1 m (i.e. coffee plantation paddock design) mounted throughout a ½" diameter (or 0.0165 m) lateral and was considered grouped in one per tree; the local head pressure losses due to emitter insertions (*hfe*) was 0.1 m; the project mean point-source discharge of 4.0 L h^{-1} was considered in principle, given that the emitter equation $q_i = 1.1134\ H_i^{0.5}$ was provided (orifice-type emitters working under full turbulent flow, according to x value); and the Blasius constants were assumed (Khf = 0.00078; m = 1.75; n = 4.75). As a project criterion, it was considered a relaxed *CVq* of 10%. The rest of variables are calculated; such as *CVHp*, lateral, losses by friction, inlet head pressure and uniformity.

The results of the simulation are presented in Table 1. The user can perform the same simulations by running the following procedure: i) calculate *CVHp* from eq. (27); ii) calculate *Hm* from eq. (23); iii) calculate L from the combination of equations (28), (29) and (30); iv) calculate H_{inl} from eq. (32); v) calculate H_i from the combination of equations (31), (34), Hf_R through eq. (30) and eq. (33); vi) calculate q_i (eq. (35)); and finally vii) calculate discharge statistic uniformity of design according to eq. (36).

Table 1: Main results of the simulation performed.

CVf	*L* (m)	H_{inl} (m)	Hf / H_{inl}	*US*	*CVHp*
2.00%	218.00	21.76	43.04%	91.18%	19.52%
4.00%	214.00	21.35	41.69%	91.55%	18.37%
6.00%	205.00	20.45	38.68%	91.55%	15.90%
8.00%	187.00	18.94	32.43%	93.98%	11.88%
10.00%	90.00	13.98	5.88%	99.12%	2.79%

It might be seen from Table 1 that required inlet head pressure tend to decrease as *CVf* increase and the same is verified the relationship between *Hf* and H_{inl}. The statistic uniformity of discharge tends to follow *CVf* increasing, as *CVHp* decreases. Interest to note that, when *CVf* = 10%, all results increase or decrease abruptly. From this value on (i.e. *CVf* > 0.1), there was no convergence for *CVHp*. This *CVf* could be considered a critical value, for example, for the scenario assumed (*So* = – 2, *D* = 0.0165 m, etc.).

The better manufacturing of emitters implies a lower *CVf* value, which allows for lengthening of laterals for a given soil slope, as shown in Figure 1 [the combination of the pairs i vs. q_i Eq. (35)], however this does not necessarily

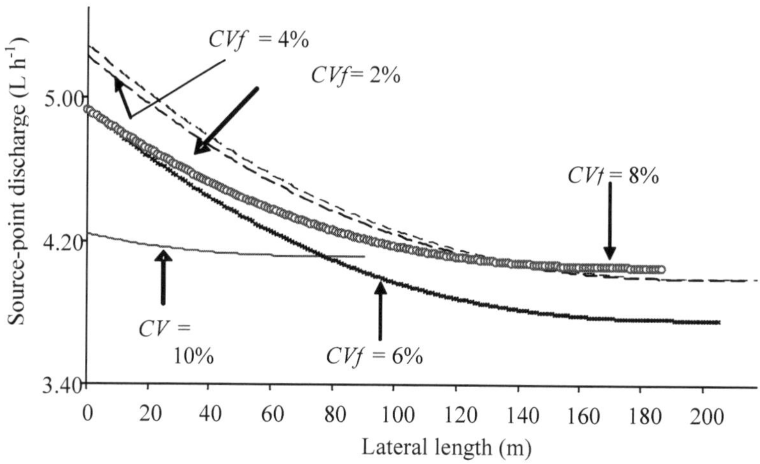

Figure 1: Point-source discharge variation throughout laterals designed according to the statistical approach.

mean that the discharge statistical uniformity throughout lateral will be more homogenous. It is possible to note that the flatter is the shape of the curve the shorter it is the simulated lateral length. Corroborating to this, it is also highlighted a decreasing tendency in statistical uniformity as emitters are more uniformly made and hence resulting in lower *CVf* values (Figure 2).

As outlined by Wu [14], the sum of squares relationship shown in the second term existing into eq. (27) indicates the effect of hydraulic design will be less significant when the emitters have high manufacturer's variations. Hence, the flatter slopes obtained from the higher *CVf*s in Figure 1, corroborated by Figure 2, reinforce such relationship. Another important characteristic to verify in Figure 1 is not only the distribution of source-point discharge values but also the limits of the corresponding head pressure values (H_i), often leading to quite distant values of q_i from the suggested mean discharge (4.0 L h^{-1}). The designed mean source-point discharge decreased from 4.36 to 4.15 L h^{-1}, as well as has happened to the emitter flow relative variation [(generally called $q_{var} = (q_{max} - q_{min})/q_{max}$], going from 24.48 to 2.91%.

From the lowest (i.e. least variable) to the highest value of *CVf*, this simple emitter attribute alteration has reduced the lateral length for more than a half although a 2% *CVf* still led to an acceptable value of micro-irrigation project uniformity (i.e. greater than 90%). The choice of a 10% *CVf* would lead to the lowest pressure head at the beginning of the lateral (13.98 m), a value 64% lower than the necessary pressure head if a 2% *CVf* emitter were selected. Thus, a 10% *CVf* emitter would result in a more economical system with respect to energy expense considered alone, but would be the shorter lateral line on the other hand. Wu [14] report that when the manufacturer's variation of selected emitters is selected to be < 10%, which is very easy to achieve, the system uniformity < 20% in CV can be achieved when the q_{var} is 30%.

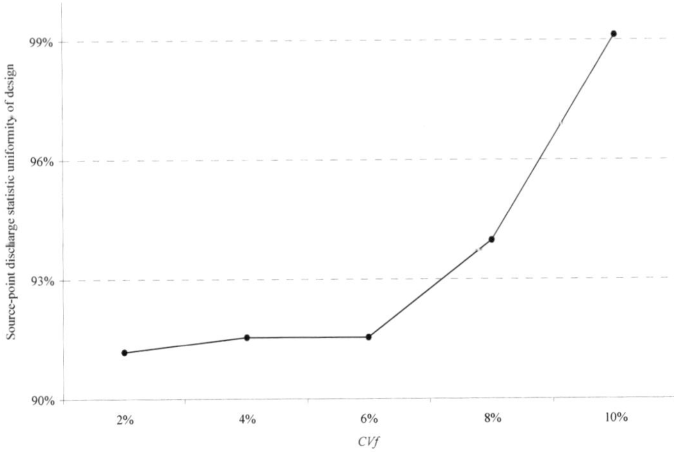

Figure 2: Variation of statistic uniformity as a function of *CVf* for a specific condition of irrigation project design.

Factors such as *CVf*, *CVH*, *CVq*, H_i and x should be carefully considered when a micro-irrigation system is designed and managed in order to ensure an application of water that is as uniform as possible. However, regular evaluation of irrigations should take place when systems are working so that they can be correctly maintained and can closely perform according to its original design (Pereira [12]). Moreover, one can also create an optimisation model to find the best trade-off between lateral length and *CVf*, thereby maximise profit, however this is not the primary goal of this work. The costs of different emitters, supposedly varying with the level of manufacturing, need to be compared as well, since it is useful to assess the profitability of an irrigation system by taking into account the fix costs involved. It is often necessary when dealing with manufacturing variation of a particular emitter to reinforce its applicability in practical situations of systems design. Therefore, it is always worthwhile to outline some crucial concepts and methodologies by bringing them in an accessible manner for educational purposes, i.e. by deducting eq. (27) in order to support that not always will be the lowest (or even highest) *CVf* the best choice.

4 Conclusions

From partial differential equations applied to error theory, it is possible to deduct the variance of discharge in order to obtain the coefficient of variation of pressure on project instead of using the deduction from Taylor Series for bidimensional variables, then to simulate lateral length in microirrigation projects. The larger the manufacturing coefficient variation, the longer will be the simulated length, which can be economically interesting until certain limits of discharge uniformity at project stage, as this performance index will be lower for longer laterals.

Appendix

List of symbols:

q	Source-point discharge [$L^3\ t^{-1}$]
K	Coefficient in emitter discharge formula [$L^{3-x}\ t^{-1}$]
x	Exponent of emitter discharge equation
H	Head pressure [L]
ρ	Correlation
δ^2	Variance
d	Deviation or differential
CVq	Coefficient of variation in discharge
CVH	Coefficient of variation of pressure heads
$CVHp$	Coefficient of variation of pressure heads at design stage
$CVK = CVf$	Coefficient of variation in discharge due to manufacture
$E(q)$	Expectation of q [$L^3\ t^{-1}$]
L	Lateral length of emitter insertion [L]
Khf	Constant in empirical equation of head loss
m	Exponent of discharge in empirical equation of head loss
n	Exponent of diameter in empirical equation of head loss
N	Number of emitters
Hf	Total head losses along a working lateral [L]
ΔZ	Elevation or level difference [L]
So	Slope [L, if divided by 100; otherwise is dimensionless]
Se	Emitter spacing [L]
hfe	Equivalent length of emitter [L]
D	Lateral internal diameter [L]
F	Christiansen' factor
CR	Relative length
US	Statistic uniformity of discharge

List of subscripts:

H	Head pressure
i	Order of a number of series
inl	inlet
K	Coefficient in emitter discharge formula
m	Mean
R	Recalculated
x	Exponent of emitter discharge equation

References

[1] Anyoji, H.; Wu. I.P. Statistical approach for drip lateral design. Transactions of ASAE, v. 30, p.187-192, 1987.

[2] ASAE, 1996. ASAE standards engineering practices data. 43 ed. MI, USA. 864p.

[3] Bralts, V.F. Field performance and evaluation. In: Nakayama, F.S.; Bucks, D.A. (Eds.) Trickle irrigation for crop production (Development in Agricultural Engineering, 9). Amsterdam: Elsevier, 1986. p.216-240.

[4] Burden, R.L.; Faiures, J.D. 2004. Numerical Analysis. 8 ed. New York: Brooks Cole, 847p.

[5] Faveta, G.M.; Botrel, T.A. Uniformity of localized irrigation systems: validation of equations. Scientia Agricola, v. 58, p.427-430, 2001.

[6] Frizzone, J.A.; Vieira, A.T.; Paz, V.P.S.; Botrel, T.A. Hydraulic characterisation of a drip tape. Revista Brasileira de Engenharia Agrícola e Ambiental, v. 2, p.278-283, 1998.

[7] Juana, L.; Losada, A.; Rodriguez-Sinobas, L.; Sánchez, R. Analytical relationships for designing rectangular drip irrigation units. Journal of Irrigation and Drainage Engineering, v. 130, p.47-59, 2004.

[8] Keller, J.; Bliesner, R.D. Sprinkler and trickle irrigation. New York: Von Nnostrand Reinhold, 1990. 582p.

[9] Lindley, D.V. Introduction to probability and statistics (Part I). Cambridge: University Press, 1965.

[10] Moore, D.S. The basis practice of statistics. New York: W. H. Freeman and Company, 1995. 680p.

[11] Peng, G.F.; Wu, I.P.; Phene, C.J. Temperature effects on drip line hydraulics. Transactions of ASAE, v. 29, p.211-215, 1986.

[12] Pereira, L.S. Higher performance through combined improvements in irrigation methods and scheduling: a discussion. Agricultural Water Management, v. 40, p.153-169, 1999.

[13] Wilcox, J.C.; Swailes, G.E. Uniformity of water distribution by some under tree orchard sprinklers. Scientific Agricultural, v. 27, p.565-583, 1947.

[14] Wu, I.P. An assessment of hydraulic design of micro-irrigation systems. Agricultural Water Management, v. 32, p.275-284, 1997.

Saving irrigation water by accounting for windbreaks

T. T. de Vries, T. A. Cochrane & A. Galtier
Department of Civil and Natural Resources Engineering,
University of Canterbury, New Zealand

Abstract

Water for irrigation in the Canterbury region of New Zealand is becoming an increasingly precious commodity, as it is in many other areas of the world. Adequate use of this resource will define the economical and environmental future of the region. Current irrigation systems, even under best management practices, over-apply water, as they do not account for spatial variability of crop water needs in fields. Over-application of water is wasteful and has environmental and economical repercussions. Water requirements are determined by crop evapotranspiration (ET). Key factors affecting ET in Canterbury are wind and solar radiation. Both of these are significantly affected by windbreaks, resulting in variability in ET and water requirements across a field. Understanding the variability in ET caused by windbreaks will enable for the correct application of water through precision irrigation systems.

A theoretical model was developed to estimate savings in irrigation by accounting for windbreaks in the Canterbury region. Windbreaks reduce evapotranspiration and therefore crops/pasture behind windbreaks needs less water than those in other parts of the field. Results for a case study in Canterbury show that windbreaks can potentially reduce the annual on-farm water consumption by 10 to 20%, while still maintaining ideal crop/pasture yields. In the short term, the application of precision irrigation systems in fields with windbreaks can have the farm level benefits of improved water use and reduced nitrogen/phosphorus leaching. In the long term this could translate directly into cost savings because of a potential decrease in energy used for irrigation (running pumps, etc.).

Keywords: irrigation modelling, evapotranspiration, windbreaks, water saving.

WIT Transactions on Ecology and the Environment, Vol 134, © 2010 WIT Press
www.witpress.com, ISSN 1743-3541 (on-line)
doi:10.2495/SI100111

1 Introduction

The Canterbury region lies in the Eastern part of the South Island of New Zealand. This area stretches from the Southern Alps in the West down to the Pacific Ocean in the East. Due to the rain shadow effect of the Southern Alps, rainfall averages only 650mm per year, with occasional long dry spells, especially during the summer. A large part of the region is dominated by the Canterbury plains, one of the prime agricultural areas in New Zealand. The region was originally mostly under dryland sheep farming and arable cropping; however, since the 1990s there has been a significant increase in intensive dairy farming (Wilson [1]). Most dairy farms in the region use direct grazing only with a stocking rate of 2.5-4 cows/ha. This practice in combination with the low rainfall means irrigation has become essential for economic viability. Large irrigation schemes have been completed and more are proposed to cope with the increased demand. This has increased the pressure on both surface and groundwater resources and water in many areas of the region is now either fully allocated or even over allocated. In addition there is an increased demand for energy (through pumping) and an increased risk of groundwater nitrate contamination associated with increased dairying and irrigation.

In addition to low rainfall and intensive grazing, another influence on irrigation requirements are the high wind speeds Canterbury experiences. To reduce the effect of the wind on crop evapotranspiration and soil erosion, many farms have established windbreaks. The development of windbreaks started in the 1850s and by the 1990s extended for a combined length of nearly 300,000 km (Price [2]). While a large number of fields have barriers on all four sides, most have at least a barrier in the East-West direction. This is to protect crops and livestock from the southwesters which, although not the prevailing wind, are the most damaging as they tend to bring storm-force winds straight from Antarctica (the temperature differential between northerlies and southerlies can be 15-20 °C and can occur is very short periods of time).

Windbreaks work by the adsorption of momentum from the wind flow, which results in a decrease of the wind velocity and turbulence (Vigiak et al. [3]). The influence of the windbreak depends on the height and porosity and extends from around 5 times the height on the windward side to 35 times the height on the lee side (Figure 1). As shown wind speeds can be reduced up to 90%, which may have a considerable influence on a crops growing in the influence zone of a windbreak.

Evapotranspiration (ET) rates of a crop are dependent on many factors such as net radiation, air temperature, vapour pressure, but also wind speed. Any change in the wind speed is likely to have an effect on the ET. In the case of a well-watered crop reduced wind speeds behind a windbreak will result in a drop in ET (Cleugh [4]). At low wind speeds this drop will be minimal, but with high wind speeds and low humidity this drop may be up to 45% (Doorenbosch and Pruitt [5]). Both conditions are a common occurrence on the Canterbury Plains during the summer.

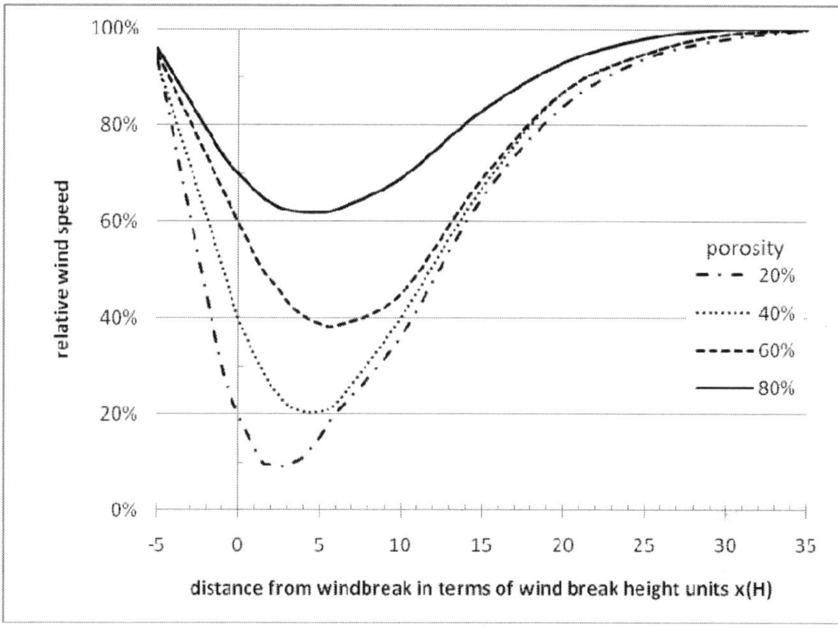

Figure 1: Reduction in wind speed behind wind breaks of different porosities (after Vigiak et al. [3]).

Windbreaks also create a shadow close to the barrier which further reduces ET. Shadows can significantly reduce the radiation received by a crop close to the wind break and thus reduce the ET. Although the wind reducing effect extends much further into the field, the shading can affect the first 20-50m. If the distance between two windbreaks is as small as 100m, as happens often in Canterbury, this may affect total ET in a field considerably, especially during the early and late season (as a result of the lower solar angle).

Accounting for a drop in ET in parts of a field has very little practical application if there is no possibility to translate this drop into a reduced irrigation water requirement and reduced water application. Until recently most irrigation systems in the Canterbury area were fixed rate centre-pivots or linear move sprinklers which offer very little opportunity to adjust application rates. The recent development of variable rate applicators combined with GPS systems has now changed this. Therefore the aim of this paper is to model and quantify the influence of windbreaks on irrigation water requirements for a typical field in the Canterbury area of New Zealand.

2 Methodology

Irrigation water requirements for typical fields with windbreaks in Canterbury were estimated by calculating actual ET for a pasture crop at various distances from a windbreak. Wind break characteristics, climatic data, ET equations, and computational assumptions are described below.

2.1 Windbreak characteristics in Canterbury

Canterbury windbreaks are characterised by single or multiple rows of Monterey pine (Pinus radiata) and Monterey cypress (Cupressus macrocarpa), similar to the "Line 1" class proposed by Vigiak et al. [3] shown in Table 2. The Canterbury windbreaks tend to be somewhat lower and more closed at the bottom than the described Line 1 class; therefore the height has been adjusted to 10m with a porosity of 0.40. The distance between two windbreaks can be as low as 100-250m, but can also reach distances of 500-1000m. For modelling purposes the orientation of the windbreak has been set to East-West. The main purpose for irrigation is pasture for dairy farms and the dominating soil type is a sandy clay loam.

2.2 Climatic data

Climatic data was obtained from the National Climate Database of New Zealand (NIWA [6]). This included hourly measurements of temperature, dew point temperature, wind speed, direct solar radiation, rainfall and sunshine hours. Evapotranspiration was modelled for three months, March, June and December in 2004 and 2008, representing average autumn/spring, winter and summer months in the region. Evapotranspiration was calculated with the ASCE standardised reference ET equation for three scenarios as described below.

Table 1: Characteristics of windbreak classes[1].

Class	Description	Height	Width	Porosity
		(m)	(m)	(-)
Hedgerow	Low, mostly hawthorn	2.1	1.1	0.25
Hedge	Higher than hedgerow, uniform vertical screen mixture of species	8.8	4.8	0.32
Line 1	Single or multiple lines of coniferous trees	15.1	9.5	0.49
Line 2	Single line of deciduous species, mostly oak	12.5	6.2	0.46
Line 3	Multiple lines of trees, deciduous or mixed	16.2	16.6	0.25
Wood	Small woods, width/height ratio >1	15.4	29.0	0.17

[1] Source: Vigiak et al. [3].

2.3 Scenarios

Three scenarios were chosen for modelling:

- **W**, wind only, evapotranspiration is affected by windbreaks through a reduction in wind speed:

$$ET = f(u, x) \tag{1}$$

where ET = reference evapotranspiration; u = wind speed; and x = distance from windbreak.

- **S**, shade only, evapotranspiration is affected by the effect of the shade from the windbreak on the net radiation:

$$ET = f(R_n, x) \tag{2}$$

where R_n = net solar radiation.

- **WS**, wind and shade, evapotranspiration is affected by both reduced wind speeds and the shade from the windbreak:

$$ET = f(u, R_n, x) \tag{3}$$

In all scenarios the evapotranspiration is influenced by the height, width and porosity of the windbreak.

2.3.1 Wind speed *u*

Vigiak et al. [3] describe the windbreak model as used by WEPS, a process based wind erosion model. This model will be used to calculate the reduction in wind speed across a field.

$$f_{xh} = 1 - \exp[-axh^2] + b\exp[-0.003(xh + c)^d] \tag{4}$$

where f_{xh} = friction velocity reduction; xh = distance from the windbreak in windbreak heights; and a, b, c and d = coefficients depending on windbreak porosity θ.

$$a = 0.008 - 0.17\theta + 0.17\theta^{1.05} \tag{5}$$
$$b = 1.35\exp(-0.5\,\theta^{0.2}) \tag{6}$$
$$c = 10(1 - 0.5\theta) \tag{7}$$
$$d = 3 - \theta \tag{8}$$

Windbreak porosity is dependent on the optical porosity, width and height

$$\theta = op + 0.02\frac{w}{h} \tag{9}$$

The friction velocity is related to the average wind speed through the following

$$U(z) = \frac{u*}{k}\ln\left(\frac{z}{z_0}\right) \tag{10}$$

where $U(z)$ = average wind speed; $u*$ = wind friction velocity; k = von Kármán constant (0.4); z = height; and z_0 = roughness height.

The friction velocity at a weather station can then be used to calculate the friction velocity on a field

$$u_*^R = u_*^{WS}\left(\frac{z_0^R}{z_0^{WS}}\right)^{0.067} \tag{11}$$

where u_*^R = friction velocity at the field; u_*^{WS} = friction velocity at the weather station; z_0^R = roughness height at the field; and z_0^{WS} = roughness height at the weather station.

2.3.2 Net radiation R_n

The shade produced by a windbreak varies with the time of day as well as with the time of year. The length of the shadow can be calculated from the angle of the sun above the horizon and the solar hour angle

$$L(h) = \frac{\cos\omega}{\tan\beta} \tag{12}$$

where $L(h)$ = the length of a shadow in units of windbreak height; ω = angle of the sun above the horizon; and β = solar hour angle.

Figure 2 shows the difference between the effect of reduced radiation as a result of an overcast day and as a result of shading by a windbreak. On an overcast day the radiation is reduced to diffuse radiation only. On March 20th, 2004, the first graph in the example, the radiation is reduced to approximately 0.4 MJ h^{-1} m^{-2} throughout the day. The second graph of April 20th show how the effect of shading reduces the radiation to around 0.1 MJ h^{-1} m^{-2}, only to rapidly rise to normal levels once the sun rises high enough. The shadow effect only occurs on sunny days, which have been defined as days with a cloudiness of 6 octa or less.

2.4 Evapotranspiration

Evapotranspiration was calculated using the ASCE standardised reference evapotranspiration equation (Allen et al. [7]). This equation is based on the ASCE Penmann-Monteith and associated equations. The equation takes into account a variety of factors including parameters affected by windbreaks such as radiation, temperature, and wind speed.

$$ET_{sz} = \frac{0.408\Delta(R_n - G) + \gamma\frac{C_n}{T+273}u_2(e_s - e_a)}{\Delta + \gamma(1 + C_d u_2)} \tag{13}$$

where ET_{sz} = standardised reference crop evapotranspiration (mm d^{-1} or mm h^{-1}); Δ = slope of saturation vapour pressure-temperature curve kPa °C^{-1}); R_n = calculated net radiation at the crop surface (MJ m^{-2} d^{-1} or MJ m^{-2} h^{-1}); G = soil heat flux density (MJ m^{-2} d^{-1} or MJ m^{-2} h^{-1}); γ = psychrometric constant

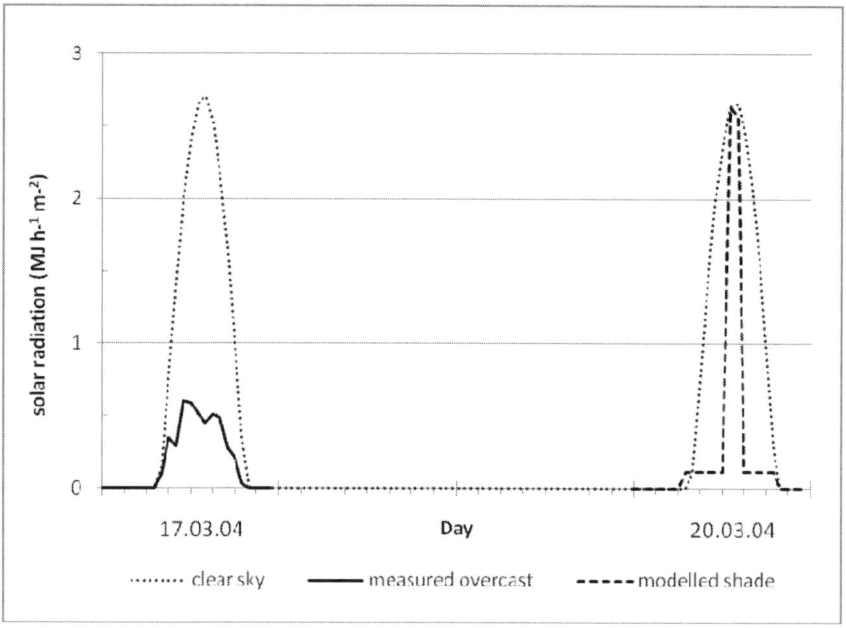

Figure 2: Comparison of (theoretical) clear sky radiation, measured radiation on an overcast day and radiation in the partially shaded area behind a wind barrier.

Table 2: Values for C_n and C_d.

Calculation time step	Short reference, ET_{os}		Tall reference crop, ET_{rs}	
	C_n	C_d	C_n	C_d
Daily	900	0.34	1600	0.38
Hourly, daytime	37	0.24	66	0.25
Hourly, night time	37	0.96	66	1.70

Source: Allen et al. [7].

(kPa °C^{-1}); C_n = numerator constant dependent on reference crop type and calculation time step (K mm s^3 Mg^{-1} d^{-1} or K mm s^3 Mg^{-1} h^{-1}); T = mean air temperature (°C); u_2 = wind speed at 2m height (m s^{-1}); e_s = saturation vapour pressure (kPa); e_a = mean actual vapour pressure (kPa); C_u = denominator constant that changes with reference crop type and calculation time step (s m^{-1}).

Eqn (13) is valid for both short reference crop evapotranspiration, ET_{os}, and tall reference crop evapotranspiration, ET_{rs}. Values for C_n and C_d are given in Table 2. As the modelled crop is pasture, ET_{os} was used.

2.5 Irrigation requirements

Reference evapotranspiration was calculated for hourly periods and summed to obtain daily values. Crop evapotranspiration was calculated from

$$ET_c = ET_{os} * K_c \tag{14}$$

where ET_c = crop evapotranspiration; and K_c = crop factor. The K_c for a well established grass pasture is 1.0.

Irrigation requirements were calculated from

$$I = \frac{(ET_c - P)(1 + LR)}{E_a} \tag{15}$$

where I = irrigation requirement; P = effective rainfall; LR = leaching requirement; and E_a = application efficiency. For the purpose of this study the leaching requirement was set to 0.1 and the application efficiency to 0.9.

3 Results and discussion

Results from modelling ET when considering the effect of wind only, shading only, and the combined effect of both are shown in Figure 3 for a single day (27/12/2004), when the wind is flowing perpendicular to the barrier. This

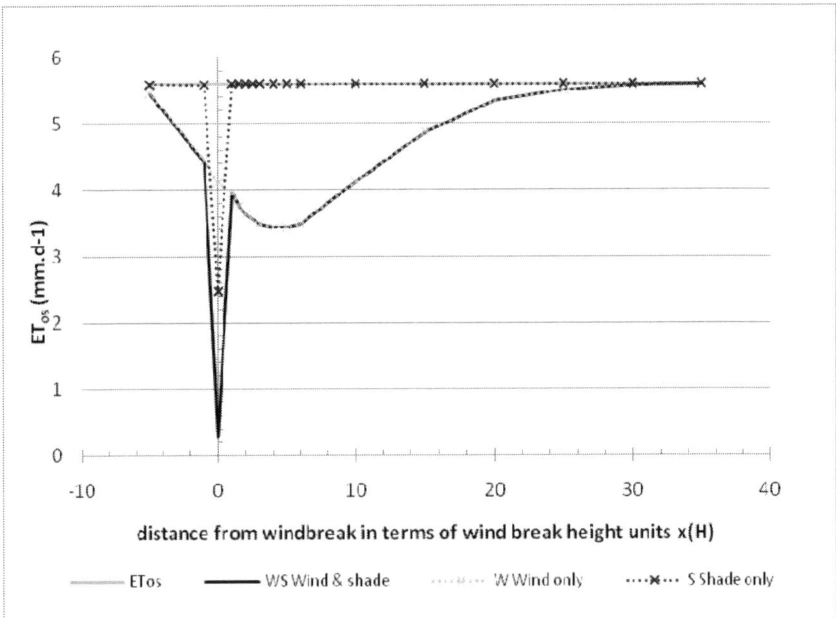

Figure 3: Reduction in evapotranspiration (ET_{os}) as a result of wind speed reduction, shade and both (27-12-2004).

example clearly illustrates that a reduction in wind due to wind barriers has a greater spatial influence on ET than changes in radiation due to shading. The effect of shading is limited to short distances from the barrier during summer, but extends up to 4 times the barrier height (H) during the winter months.

Results of modelling changes in ET (including windbreak induced changes in both wind and shade) over the months of December, March and June are shown in Figure 4. The percent reduction in ET at a distance of 5 times the height of the barrier is 26%, 22%, and 50% for the months of December, March, and June respectively. Higher wind speeds and more overcast days contributed to larger reductions in ET for the month of December compared to the month of March.

Using results from the changes in ET due to wind and shading, irrigation requirements were estimated for a pasture crop for the months of December, March and June of 2004 in Canterbury (Table 3 and Figure 5). Significant savings in water are possible if precision irrigation systems are used to apply required amounts of water behind barrier. For a 200 by 200 meter field this would translate to a yearly savings of over 16% of total irrigation water requirements.

Changes in temperature from barriers were also considered; however, preliminary simulations show that the changes in ET due to changes in temperature behind barriers are relatively small compared to the direct effects of

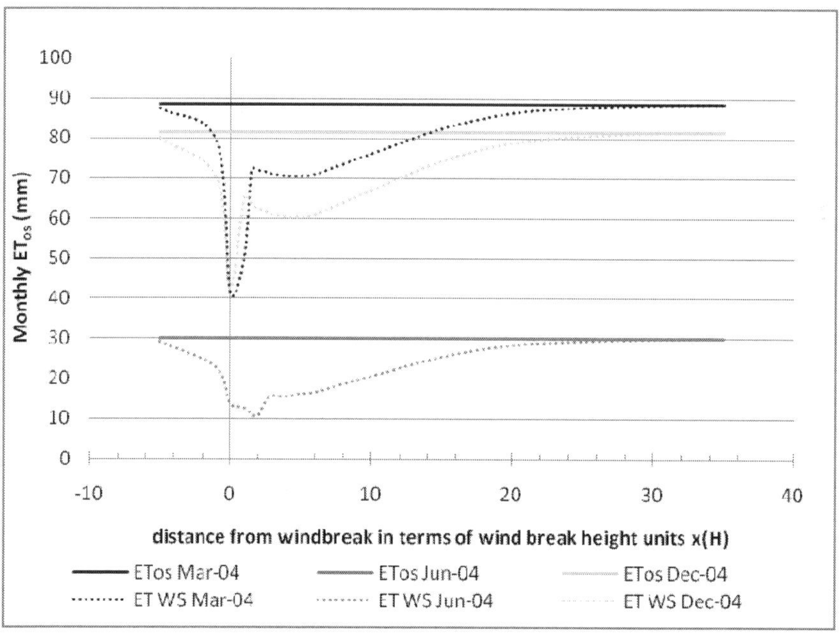

Figure 4: Modelled ET for March, June and December, 2004 at distance x(H) behind a 10 metre windbreak with shade and wind reductions.

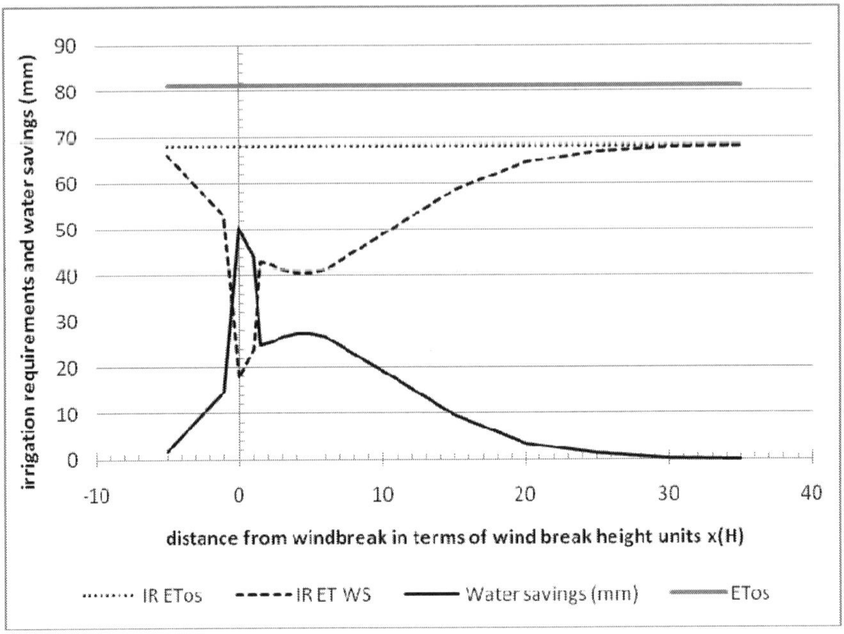

Figure 5: Comparison irrigation requirements, December 2004.

Table 3: Water saving comparison across months.

	water savings(mm)			water savings (%)		
x(H)	Mar-04	Jun-04	Dec-04	Mar-04	Jun-04	Dec-04
-5	1.18	0.88	1.54	1	3	2
-1	9.65	7.25	11.37	11	24	14
0	47.08	16.14	38.87	53	54	48
1	38.73	17.25	15.90	44	57	20
2	16.62	19.43	18.95	19	65	23
3	17.60	14.37	20.41	20	48	25
4	18.08	14.55	20.97	20	48	26
5	18.08	13.82	20.97	20	46	26
10	12.39	9.36	14.33	14	31	18
15	6.10	4.56	7.02	7	15	9
20	2.24	1.66	2.57	3	6	3
25	0.84	0.62	0.96	1	2	1
30	0.17	0.12	0.19	0	0	0
35	0.00	0.00	0.00	0	0	0

wind reduction or shading. Furthermore, it is inconclusive whether net temperatures would increase or decrease due to the indirect reduction of wind or increased shading.

4 Conclusions

Significant savings can be made by incorporating a windbreak model in the calculation of irrigation requirements in the Canterbury region in New Zealand. These savings are dependent on the field and windbreak configuration and the position in a field, but can amount up to 50% in summer and even more during the winter months.

Average annual savings of 10-20% may be possible if this model were to be used in combination with precision irrigation systems. In Canterbury this translates to significant savings, if the model was applied to all fields with windbreaks.

Work is currently underway to extend this model into a 2 dimensional model that accounts for wind direction and multiple barriers.

References

[1] Wilson, J., Canterbury region - Agriculture after 1900. Te Ara - the Encyclopedia of New Zealand. [Online] 2 March 2009. http://www.TeAra.govt.nz/en/canterbury-region/8.

[2] Price, L.W., Hedges and shelterbelts on the Canterbury Plains, New Zealand: transformation of an antipodean landscape, *Annals of the Association of American Geographers*, 83(1), pp. 119-140, 1993.

[3] Vigiak, O., Sterk, G., Warren, A. & Hagen, L.J., Spatial modelling of wind speed around windbreaks. *CATENA*, 52(3-4), pp. 273-288, 2003

[4] Cleugh, H.A., Effects of windbreaks on airflow, microclimates and crop yields. *Agroforestry Systems*, 41(1), pp. 55-84, 1998.

[5] Doorenbosch, J. & Pruitt, W.O., *Crop water requirements*, Food and Agricultural Organisation of the United Nations (FAO): Rome, 144 p., 1992.

[6] NIWA, *The National Climate Database*, http://cliflo.niwa.co.nz/

[7] Allen, R. G., Walter, I.A., Elliott, R.L., Howell, T.A., Itenfisu, D. and Jensen, M., *The ASCE Standardized Reference Evapotranspiration Equation*. American Society of Civil Engineers (ASCE): Reston, (2005).

Rainfall data calculation using Artificial Neural Networks and adaptive neuro-fuzzy inference systems

L. Mpallas, C. Tzimopoulos & C. Evangelidis
Department of Rural and Surveying Engineering, Aristotle University of Thessaloniki, Greece

Abstract

Missing rainfall data is a frequent problem in every hydrological or water management study. Conventional methods are usually based on statistics and the missing rainfall data are estimated using neighbouring stations. The same procedure is implemented here using two models based on artificial intelligence techniques. A fuzzy logic and an Artificial Neural Network model are used to predict rainfall heights of two meteorological stations. A number of model parameters were modified during the calibration, in order to achieve the best performance. The comparison between observed and estimated values showed that both models perform satisfactorily in the prediction of rainfall data.
Keywords: fuzzy logic, rainfall, neural networks, hydrology.

1 Artificial intelligence

Artificial intelligence includes a number of methods that mimic human reasoning. These methods can be modelled and applied by a computer. The most common ones are Artificial Neural Networks (ANN), Genetic Algorithms (GA) and fuzzy logic theory.

Artificial intelligence is an alternative and also effective method in studying and modelling hydrological systems. They capture their ambiguous and inaccurate nature, even when non-linearities exist. These models are extremely effective when the knowledge of the system is insufficient or when the underlying procedures are not fully understood.

WIT Transactions on Ecology and the Environment, Vol 134, © 2010 WIT Press
www.witpress.com, ISSN 1743-3541 (on-line)
doi:10.2495/SI100121

Over the last decades, models based on ANN are applied into hydrology and water management problems. They have been used in river flow prediction [1–3], in reservoir management [4], in lake fluctuation [5] etc.

Fuzzy logic model are also widely applied in groundwater movement [6], in hydrologic cycle description [7], in reservoir management [8 11], in prediction [12–15] etc.

2 Fuzzy inference systems

In Fuzzy Logic, elements are possible to belong partially to a set. Fuzzy sets are a generalization of the classical sets since they don't have clearly defined boundaries. If X is a collection of objects denoted generally by x, then a fuzzy set Ã in X is a set of ordered pairs:

$$\tilde{A} = \left\{ \left(x, \mu_{\tilde{A}}(x) \right) \middle| x \in X \right\} \tag{1}$$

where $\mu_{\tilde{A}}(x)$ is called the membership function or grade of membership of x in Ã. The degree to which an element belongs to a fuzzy set is shown by the membership function, which takes values from the [0, 1] interval.

A fuzzy rule is composed of explanatory variables (fuzzy premises) connected together with operators and an action entailed, which gives the output (fuzzy response). These operators associate the input variables and their combination defines the result. The difference between the fuzzy rules and the rules corresponding to common logic is that the first allow partial and simultaneous applicability. Partial applicability means that the membership functions of the input values belong to the [0, 1] interval.

A collection of fuzzy IF-THEN rules constitutes a fuzzy rule-based system or a fuzzy inference system. First the input variables, which are usually crisp, are being transformed into fuzzy ones, by applying a membership function. This procedure denotes the degree to which an input value belongs to a fuzzy set. The most common membership function is the triangular one.

The fuzzy premises are connected using operators (AND, OR and rarely XOR). AND is the operator which is most commonly used, corresponding to the intersection of the classical sets. There are many ways to interpret the AND operator, defined by a great variety of t-norms, as they have been proposed by many researchers [16]. The most common ones are the minimum method and the algebraic product. The implementation of a t-norm results the Degree of Fulfilment (DOF) of a rule, which also takes values from the [0, 1] interval.

The DOF shows the applicability of each rule for every vector of premises. Since the fuzzy premises are being fulfilled to some degree, the fuzzy consequences should be fulfilled to the same degree. The fuzzy responses of each rule are aggregated and finally a combined fuzzy response is given.

In all cases, for practical reasons, we need to replace the resulting fuzzy set with a crisp number, usually in order to take a control action. This procedure is called defuzzification.

2.1 ANFIS (Adaptive Neuro Fuzzy Inference System)

ANFIS is a technique for constructing fuzzy inference systems and was proposed by Jang [17]. It works like neural networks and it is able to learn from a data set the membership function parameters in order to capture the system behaviour. These parameters are being adjusted during the training process in order to obtain the IF-THEN rules. The data set consists of input and output values. A gradient vector facilitates the computation of these parameters. Once the gradient vector is obtained, any of the several available optimization routines could be applied in order to adjust the parameters and reduce the model error.

Suppose a rule system consisting of two fuzzy premises x and y and a fuzzy response. The rule system is constituted by two fuzzy "IF-THEN" rules:

Rule 1: If x is A1 and y is B1 Then f1 = p1x + q1y + r1

Rule 2: If x is A2 and y is B2 Then f2 = p2x + q2y + r2

Layer 1: Every node has a membership function:

$$O_{i,1} = \mu_{A_i}(x) \tag{2}$$

where x is a crisp input, Ai a linguistic variable and μAi the membership function. If a bell-shaped function is applied then:

$$\mu_{A_i}(x) = \frac{1}{1 + \left[\left(\frac{x - c_i}{a_i}\right)^2\right] b_i} \tag{3}$$

where $\{a_i, b_i, c_i\}$ are function parameters that change the function shape.

Layer 2: In the rule nodes the AND/OR operator is applied to get one output representing the results of the antecedent for a fuzzy rule, which is, firing strength. It denotes the degrees by which the antecedent part of the rule is satisfied and it indicates the shape of the output function for that rule. So the outputs of the second layer, called by firing strengths O2,i, are the products of the corresponding degrees obtaining from layer 1, named as w in below:

$$O_{i,2} = w_i = \mu_{A_i}(x) \times \mu_{B_i}(y), \quad \mathrm{i} = 1{,}2 \tag{4}$$

Layer 3: The main target is to compute the ratio of each rule's firing strength to the sum of all rules' firing strength. The firing strength in average nodes (layer 3) is normalized.

$$O_{i,3} = \overline{w_i} = \frac{w_i}{w_1 + w_2}, \quad \mathrm{i} = 1{,}2 \tag{5}$$

Layer 4: In the consequent nodes (layer 4), the contribution of each ith rule toward the total output or the model output and/or the function defined as below is calculated.

$$O_{i,4} = \overline{w_i} f_i = \overline{w_i}(p_i x + q_i y + r_i) \tag{6}$$

where $\{p_i, q_i, r_i\}$ the parameter set in the consequent function.

Layer 5: Includes the defuzzification procedure.

3 Artificial Neural Networks

A neural network is an interconnected assembly of simple processing elements, *units* or *nodes*, whose functionality is based on the actual neurons. The processing ability of the network is stored in the connection strengths, or weights, obtained by a process of adaptation from a set of training patterns.

Human neurons communicate via electrical signals that are short-lived impulses in the voltage of the cell wall or membrane. The interneuron connections are mediated by electrochemical junctions called *synapses*, which are located on branches of the cell referred to as *dendrites*. Each neuron typically receives many thousands of connections from other neurons and is therefore constantly receiving a multitude of incoming signals, which eventually reach the cell body. They are summed together and if the resulting signal exceeds some threshold, then the neuron will "fire" or generate a voltage impulse in response. This is transmitted to other neurons via a branching fibre known as the axon.

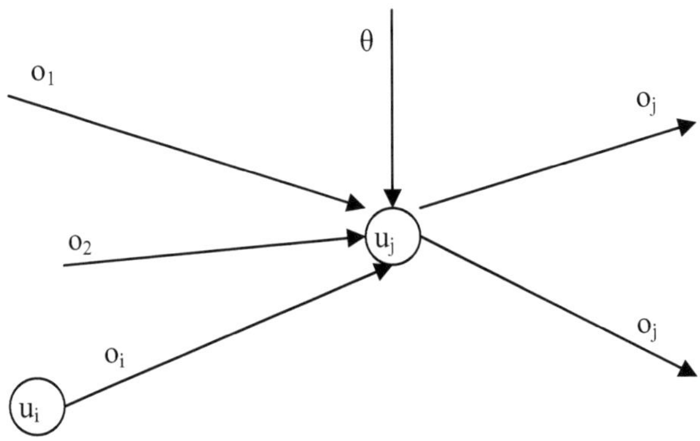

Figure 1: Artificial Neural Network (general form).

Synapses are modelled by a single number or weight so that each input is multiplied by a weight, before being sent to the equivalent of the cell body. Every unit receives signals from more than one units and the signal is sent to another one. The weighted signals are summed together by simple arithmetic addition to supply node activation. The activation is compared with a threshold. If the activation exceeds the threshold, the unit produces an output (conventionally "1"), otherwise it outputs zero. The activation or not of the node depends on a function fj, usually a sigmoid one. The sigmoid function is given by the following equation:

$$f_j = \frac{1}{1+e^{(-v_j)}} \tag{7}$$

where v a parameter that defines the exact shape.

The whole system works in parallel. Each neurons output is equal to the summed weighted input signals with a constant added. Each neurons output is an input to the next neurons until the output layer [18].

One type of network is shown in Fig. 2 where the nodes are arranged in a layered structure. Usually a neuron is connected only to the neurons of the next layer. When there are multiple layers then the intermediate ones are called hidden. This feed forward structure is only one of several available and is typically used to place an input pattern into one of several classes according to the resulting pattern of outputs [19].

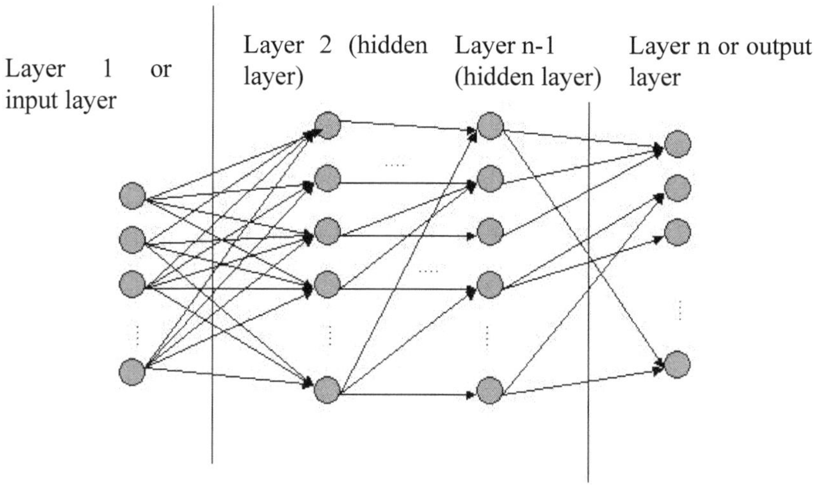

Figure 2: Typical Artificial Neural Network.

3.1 Training

In real neurons the synaptic strengths may, under certain circumstances, be modified so that the behaviour of each neuron can change or adapt to its particular stimulus input. In artificial neurons the equivalent of this is the modification of the weight values. In terms of processing information, there are no computer programs running here. The network knowledge is stored in its weights, which evolve by a process of adaptation to a stimulus from a set of pattern examples. In one training paradigm called supervised learning, an input pattern is presented to the net and its response then compared with a target output. In terms of our previous letter recognition example, an "A", say, may be input and the network output compared with the classification code for A. The difference between the two patterns of output then determines how the weights are altered. Each particular recipe for change constitutes a learning rule, details of which form a substantial part of subsequent chapters. When the required weight updates have been made another pattern is presented, the output compared with the target, and new changes made. This sequence of events is repeated iteratively many times until the network's behaviour converges so that its response to each pattern is close to the corresponding target. The process as a whole, including any ordering of pattern presentation, criteria for terminating the process, etc., constitutes the training algorithm.

3.2 Levenberg-Marquardt algorithm

Levenberg-Marquardt algorithm is the combination between Gauss - Newton method (for optimization problems) and the Gradient Descent method. It was designed to approach second-order training speed without having to compute the Hessian matrix. When the performance function has the form of a sum of squares (as is typical in training feedforward networks), then the Hessian matrix is:

$$H = J^T J \tag{8}$$

and the gradient can be computed as

$$g = J^T e \tag{9}$$

where J is the Jacobian matrix that contains first derivatives of the network errors with respect to the weights and biases, and e is a vector of network errors. The Jacobian matrix can be computed through a standard back propagation technique that is much less complex than computing the Hessian matrix. The Levenberg-Marquardt algorithm uses this approximation to the Hessian matrix in the following Newton-like update:

$$x_{k+1} = x_k - \left[J^T J + \mu I\right]^{-1} J^T e \tag{10}$$

When the scalar μ is zero, this is just Newton's method, using the approximate Hessian matrix. When μ is large, this becomes gradient descent with a small step size. Newton's method is faster and more accurate near a minimum error, so the aim is to shift toward Newton's method as quickly as possible. Thus, μ is decreased after each successful step (reduction in performance function) and is increased only when a tentative step would increase the performance function [20].

4 Application

Rainfall itself is an extremely time and location varying phenomenon. These data are usually obtained from meteorological stations, satellites or radar. The most common problem of meteorological stations is the fact that the data is usually incomplete. This is due to station malfunction, observer's negligence, etc. In these cases the estimation of missing values is implemented based on neighbouring stations. The aim of this paper is to predict the missing rainfall data of two stations, using fuzzy logic and Artificial Neural Network theory. The fuzzy inference model is trained using ANFIS method and the ANN is trained using Levenberg-Marquardt algorithm.

4.1 Meteorological stations

In this project there were four stations under examination, all located in Strymonas hydrological basin, in northern Greece. Serres and Achladochori meteorological stations are used as inputs for both fuzzy logic and ANN model. N. Zichni and Nigrita meteorological stations are used as outputs. Two runs were finally implemented for both models. In the first case the available rainfall data heights of Serres and Achladochori station were used as inputs and the corresponding data of N. Zichni were used as outputs. In the second case input data were kept the same, while Nigrita rainfall data were used as outputs. The difference between these two cases is the fact that the N. Zichni station data are more linearly correlated to the data of the input stations, than Nigrita station. The aim was to examine the influence of correlation between input and output data to model performance. The meteorological stations are:

Table 1: Meteorological stations.

Station	Altitude (m)	Available data
Serres	34	1966-today
Achladochori	485	1980-2003
N. Zichni	280	1980-2003
Nigrita	111	1980-2003

For all cases the mean monthly values were used. Part of them was used for training and the rest for validation.

4.2 ANFIS

Fuzzy inference models are based on "If-Then" rules, where "If" refers to the fuzzy premises vector. Fuzzy premises are Serres and Achladochori rainfall heights. The fuzzy response vector consists of N. Zichni and Nigrita rainfall heights and refers to the "Then" part of the rule. The rule system was trained using ANFIS, where the parameters of membership function was adapted using an optimization method. In this project a hybrid optimization system was used, which is included in MATLAB.

The training set consists of mean monthly rainfall heights from years 1980 to 1995 (192 values). Validation set was constructed using the mean monthly rainfall heights from years 1996 to 2003 (96 values).

Fuzzy logic model performance was tested when two parameters were changed. These were the type and the number of membership function utilized in dividing the domain of fuzzy premises. It must be noted that the influence of the overlap between membership functions on the results, was also investigated. We concluded that the overlap does not influence the final result significantly and finally 50% overlap between fuzzy premises was selected.

4.3 Membership function number

The number of the membership functions utilized in order to divide the fuzzy premises domain, is a parameter that affects fuzzy logic model performance significantly. Usually a trial and error method is implemented, where the number of the membership functions with the smallest model error is sought. The model was tested using from 2 to 5 membership functions for the division of the domain of Serres and Achladochori meteorological stations. It must be noted that high number of membership functions results to a high rule number. When the rule number is high then the model overfits. Finally 2 membership functions were used for model runs.

4.4 Membership function type

Fuzzy logic models are flexible in the selection of membership function type. The aim is to find the type of membership function that results the best model performance and models the system sufficiently. The model was tested using triangular, bell-shaped and Gaussian membership function. Finally the Gaussian membership function was selected although model performance was not significantly better. Gaussian membership function is given by eqn (11):

$$\mu_A(x) = e^{-(x-c)^2/2\sigma^2} \tag{11}$$

where c the point where membership function equals unit.

4.5 ANN model

The performance of the ANN model was examined in respect to the number of hidden layers when implementing the Levenberg-Marquardt algorithm. The choice of layer number was implemented through a trial and error procedure and affected model performance directly. The model reproduced the rainfall heights of N. Zichni and Nigrita stations satisfactory with the use of five hidden layers. The training set consisted of the mean monthly rainfall data from years 1980 to 1995 and validation set the data from 1996 to 2003.

4.6 Model evaluation

Model evaluation is the procedure where the results derived from both models are compared to the observed values. This is carried out for both training and validation set. Model performance is measured using reduced mean square error and coefficient of determination (R^2). Reduced mean square error is:

$$\textit{reduced}\ \mathrm{MSE} = \frac{1}{n}\sum_{i=1}^{n}\left(\frac{Y_{obs} - Y_{calc}}{Y_{obs}}\right)^2 \tag{12}$$

where Yobs the observed and Ycalc the calculated values.

5 Results

Model performance was measured for both training and validation sets using reduced mean square error and coefficient of determination (R^2) between calculated and observed values. The results obtained from model execution are shown in tables 2 and 3 for N. Zichni and Nigrita meteorological stations respectively. It must be noted that a high number of runs was performed for both

Table 2: Model performance for calculating N. Zichni station rainfall heights.

Method	Set	R^2	Reduced MSE
ANFIS	Training	0,7097	0,270025
"	Validation	0,6775	0,311568
ANN	Training	0,6469	0,307301
"	Validation	0,7069	0,297889

Table 3: Model performance for calculating Nigrita station rainfall heights.

Method	Set	R^2	Reduced MSE
ANFIS	Training	0,7352	0,261991
"	Validation	0,761	0,298543
ANN	Training	0,6769	0,302348
"	Validation	0,7798	0,267172

models in order to achieve the best performance. The results shown in tables 2 and 3 are the ones obtained with the parameter selection that performed better. The results obtained from model execution are also shown in fig. 3 and 4 graphically.

The best performance for the ANFIS model was achieved with the use of two membership functions for the division of the domain of the fuzzy premises. This means that the rule system consisted of four fuzzy rules. When the rule number increased, then the model over fitted. This means that model performance was high for the training and poor for the validation set. In this case the model is not suggested for prediction because it loses its ability to generalize. For this reason the purpose was to find the proper rule number in order to achieve similar model performance for the training and the validation sets.

Three different membership function types were tested to divide the domains of Serres and Achladochori rainfall heights. The fuzzy logic model was finally constructed with the use of Gaussian membership functions. However it must be noted that the difference in model performance was not noticeable between this type of membership function and the other two types.

Concerning the ANN model, the effect of the hidden layer number was examined in model performance. Several model run were performed in order to find the number of hidden layers which result to the best model performance. Five hidden layers were finally used. It must be noted that the hidden layer increase resulted in model over fitting, just like the fuzzy logic model. The best result was achieved when model performance was balanced between two sets.

Both fuzzy logic and ANN model reproduced the rainfall time series satisfactory. Model performance was high and both models are suggested for prediction. However they show some weakness in achieving extreme values.

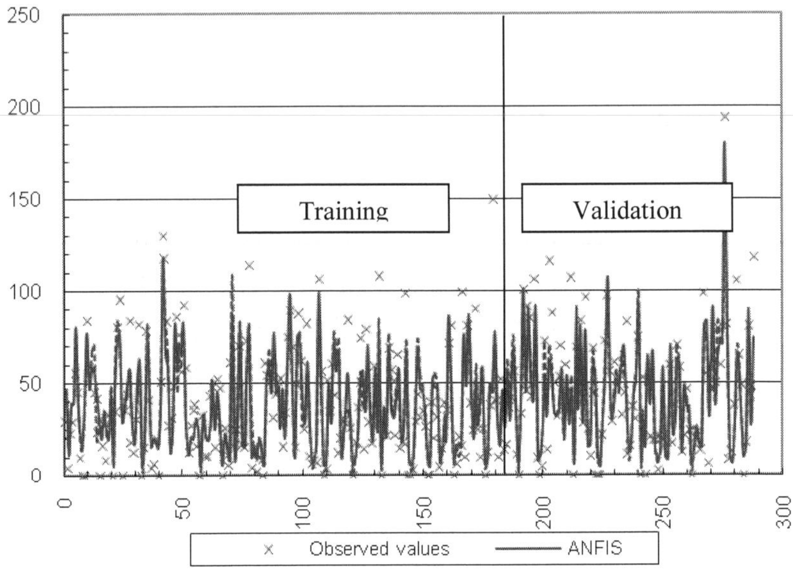

Figure 3: Mean monthly rainfall heights in N. Zichni.

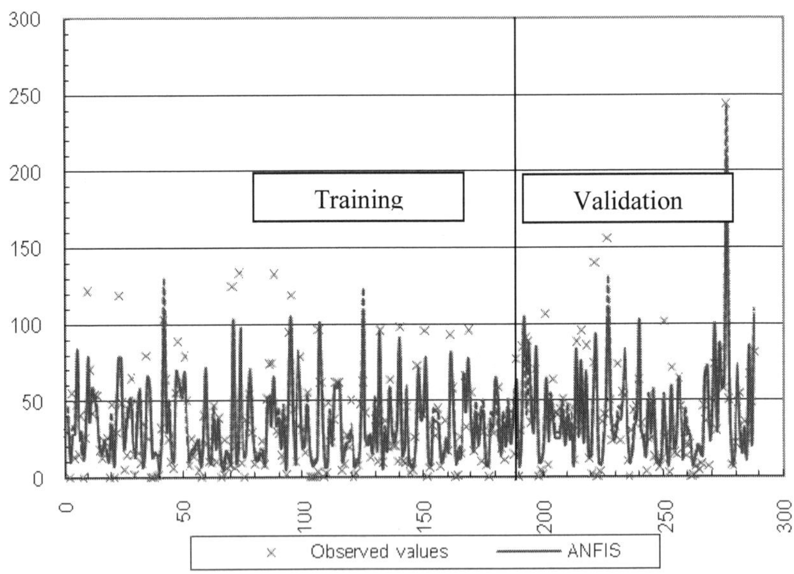

Figure 4: Mean monthly rainfall heights in Nigrita station.

Fuzzy logic model performed better in calculating rainfall heights for N. Zichni and Nigrita meteorological stations in comparison with the ANN model. The first one is much more flexible in modelling because of the greater parameter number that can be modified. The ANN is working much more like a "black box", because the final result is extracted with a less transparent way.

Both models performed better in approximating Nigrita's rainfall heights. We believe that is due to the fact input and output data are more correlated than in the case of N. Zichni station.

Acknowledgement

This research was funded by State Scholarship Foundation.

References

[1] Aquil, M., Kita I. & Yano, A., Nishiyama, S., *Neural Networks for Real Time Catchment Flow Modelling and Prediction*, Water Resources Management. **21**:1781–1796, 2007

[2] Gopacumar, R., Takara, K. & James, E., *Hydrologic Data Exploration and River Flow Forecasting of a Humid Tropical River Basin Using Artificial Neural Networks.* Water Resources Management **21**, 2007

[3] Firat, M. & Turan, E., *Monthly river flow forecasting by an adaptive neuro-fuzzy inference system*. Water and environment journal., 2009

[4] Antar, M., Elassiouti, I. & Alam M., *Rainfall-runoff modelling using artificial neural networks technique: a Blue Nile catchment case study.* Hydrological Processes **20**, 2006

[5] Altunkaynak, A., *Forecasting surface water level fluctuations of lake van by artificial neural networks*, Water Resources Management **21**:399–408, 2007

[6] Bárdossy, A & Disse, M., *Fuzzy Rule-Based Models for Infiltration.* Water Resources **29 (2)**: 373-382, 1993

[7] Bárdossy, A., The use of fuzzy rules for the description of elements of the hydrological cycle. Ecological Modeling **85**: 59-65, 1996

[8] Russell, S & Campbell, P., *Reservoir Operating Rules with Fuzzy Programming*. Journal of Water Resources Planning and Management. **122 (3)**: 165-170., 1996

[9] Shrestha, B, Duckstein, L & Stakhiv E., *Fuzzy rule-based modelling of reservoir operation.* Journal of Water Resources Planning and Management **122(4)**: 262-269., 1996

[10] Panigrahi, D & Mujumdar, P., *Reservoir Operation Modelling with Fuzzy Logic.* Water Resources Management. **14**: 89-109, 2000

[11] Chang, Li-Chiu & Chang, Fi-John, *Intelligent control for modelling of real-time reservoir operation.* Hydrological Processes **15**: 1621-1634, 2001

[12] Pongracz, R, Bogardi, I & Duckstein L., *Application of fuzzy rule-based modeling technique to regional drought*. Journal of Hydrology. **224**: 100-114, 1999

[13] Abebe, A J, Solomatine, D P & Venneker R. G. W., *Application of adaptive fuzzy rule-based models for reconstruction of missing precipitation events.* Hydrological Sciences Journal **45 (3)**: 425-436, 2000

[14] Mahabir, C, Hicks, F. E & Fayek R., *Application of fuzzy logic to forecast seasonal runoff.* Hydrological Processes. **17**: 3749-3762, 2003

[15] Nayak, P, Sudheer, K. & Ramasastri K., *Fuzzy computing based rainfall-runoff model for real time flood forecasting.* Hydrological Processes. **19**: 955-968, 2005

[16] Zimmermann, H. J., *Fuzzy Set Theory and its Applications*, Kluwer Academic Publishers: Boston, 2005

[17] Jang, J-SR., *ANFIS: Adaptive-Network-based Fuzzy Inference Systems.* IEEE Transactions on System Man and Cybernetics. **23 (3)**: 665-685, 1993

[18] Martin, JD, Morton, YT. & Zhou Q., *Neural network development for the forecasting of upper atmosphere parameter distributions.* Advances in Space Research. **36: no.12.** 2480-2485, 2005

[19] Gurney, K., *An introduction to Neural Networks*, CRC Press, Boca Raton, 1997

[20] Mathworks, *Neural Network Toolbox 6, Users guide.* MATLAB, 2007.

Design for an "engineered sinkhole" to improve, recharge and reduce evapotranspiration in an upstream flood control structure

Y. Leventeli[1], T. Halihan[2], M. Dailey[2], K. Hurt[3] & W. Kellogg[3]
[1]*Faculty of Engineering, Akdeniz University, Turkey*
[2]*School of Geology, Oklahoma State University, USA*
[3]*Chickasaw Nation, USA*

Abstract

Freshwater resource shortages have become a significant problem in the world due to population growth, rural to urban migration, climate change, and flood control. When water is stored above ground, mass amounts of water are lost due to evapotranspiration. In order to curb the freshwater resource shortage we propose the following solutions: 1) preventing some storm water from becoming runoff, 2) keeping water clean, 3) storing water underground, and 4) using water more intelligently.

Flood control structures, thousands in the United States alone, have been utilized around the world to reduce peak flow to streams in order to avoid property damage. However, these structures store the excess water with little benefit, reduce recharge, and increase evapotranspiration.

Artificial recharge is the process of augmenting groundwater through the intervention of man's activities- either intentional or unintentional. The goal in artificially recharging water to an aquifer is to increase the long term stability of a clean water supply to both people and the environment. Water stored in aquifers is not subjected to the evaporative losses of surface storage areas (e.g. lakes and ponds). One of the advantages of storing water in aquifers is that the risks associated with anthropogenic or natural contamination are reduced and environmental and surface impacts are mitigated.

This research presents a design to integrate flood control measures with artificial recharge structures in order to provide better flood control and improve

WIT Transactions on Ecology and the Environment, Vol 134, © 2010 WIT Press
www.witpress.com, ISSN 1743-3541 (on-line)
doi:10.2495/SI100131

recharge without sacrificing water quality. This study evaluated alternatives for artificial recharge in terms of environment, safety, and economy for upstream flood control dam #37 in southern Oklahoma. An "engineered sinkhole" is proposed that will maintain flood control while allowing artificial recharge to proceed without sacrificing water quality.

Keywords: artificial recharge, engineered sinkhole, Oklahoma, upstream flood control structure, Arbuckle-Simpson aquifer.

1 Introduction

In order to increase the supply of groundwater, artificial recharge has been employed using many different techniques in many different areas around the world. Artificial recharge may be defined as "augmenting the natural movement of surface water into underground formations by some methods of construction, by spreading of water, or by artificially changing natural conditions" [1]. The factors governing the choice of site specific artificial recharge methods include climate, hydrogeology, topography, source water availability and quality, operational and management issues, regulatory controls and environmental and socio-economic considerations with topography, climate, and hydrogeology being the major concerns.

The Wingard Ranch study area, in south-central Oklahoma between the small communities of Fitzhugh and Roff, is used primarily for cattle grazing. The Blue River, which is ephemeral along this reach, runs through the property (Figure 1).

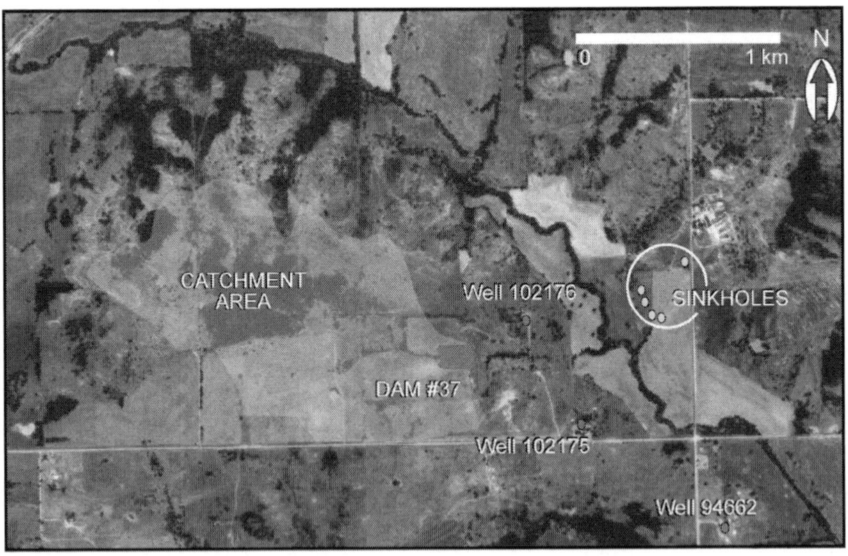

Figure 1. Aerial photograph of the study site with the location of dam #37 and its catchment area illustrated. Riparian zones darker due to high tree density [aerial photograph from 5].

Numerous sinkholes are located on the east side of the Blue River. These sinkholes serve as natural recharge features for the aquifer. Upstream Flood Control Dam #37 was built as a watershed dam in 2000 on the Wingard Ranch and occupies 1.6 hectares at the conservation pool and has a flood storage capacity of approximately 185,000 m^3 [2]. The watershed drains approximately 180 hectares [2]. With an average rainfall of 1.01 m [3] and assuming a runoff coefficient of 0.2, this watershed is capable of generating 287,000 m^3 of runoff. Overflow from the dam is discharged through an outlet structure into a plunge pool where the water eventually drains into the Blue River located approximately 650 m to the east.

While this structure and pool is constructed on a sandstone formation that has shale interbeds that limit vertical permeability, the nearby sinkholes allow rapid recharge into the underlying carbonate formation. In this study, an "engineered sinkhole" design is evaluated as an artificial recharge method in terms of environment safety, and economy for the Wingard Ranch and its upstream flood control structure.

2 Climate and topography

The average temperature is 16.7° Celsius. The average annual precipitation is 1.01 m [3]. The majority of precipitation happens during the spring months followed by dry summers. The annual lake evaporation averages 1.4 m with 70% occurring from May to October [4].

The elevation ranges between 335 m and 396 m in the study area, fig. 1. The northeastern portion of the study area has relatively rough topography while the area around the watershed dam has a relatively flat landscape (less than 3% slope). The Blue River and its tributaries trend northwest to southeast. The elevation of Blue River is 354 m in the northwest and 344 m in the southeast in the study area. The tributary of the Blue River, which the dam is located on, starts from 372 m and combines with the main stream at 349 m. The dam's overflow elevation is 354 m [2].

3 Hydrogeology

The geological and structural setting has an important influence on site selection for artificial recharge. The overlying formations, with their thickness and aerial extent, affect the input to aquifers and the methods used to recharge the aquifer. The spatial extents of the formations are shown in figure 2.

According to Fairchild et al. [6], the study site is located over the Arbuckle-Simpson aquifer. There are two main groups in the study area, the Arbuckle and the Simpson Group. The Simpson Group in the area consists mainly of a gray to tan limestone with a greenish-gray shale and fine to medium grained, brown sandstone with its depth ranging between 1 m and 76 m. The Arbuckle Group in the area is mainly a gray to tan limestone, which grades eastward into a tan dolomite with some stringers of shale and sandstone, with a thickness of more than 300 m. There is a fault on the north side of the Simpson Group in the study area.

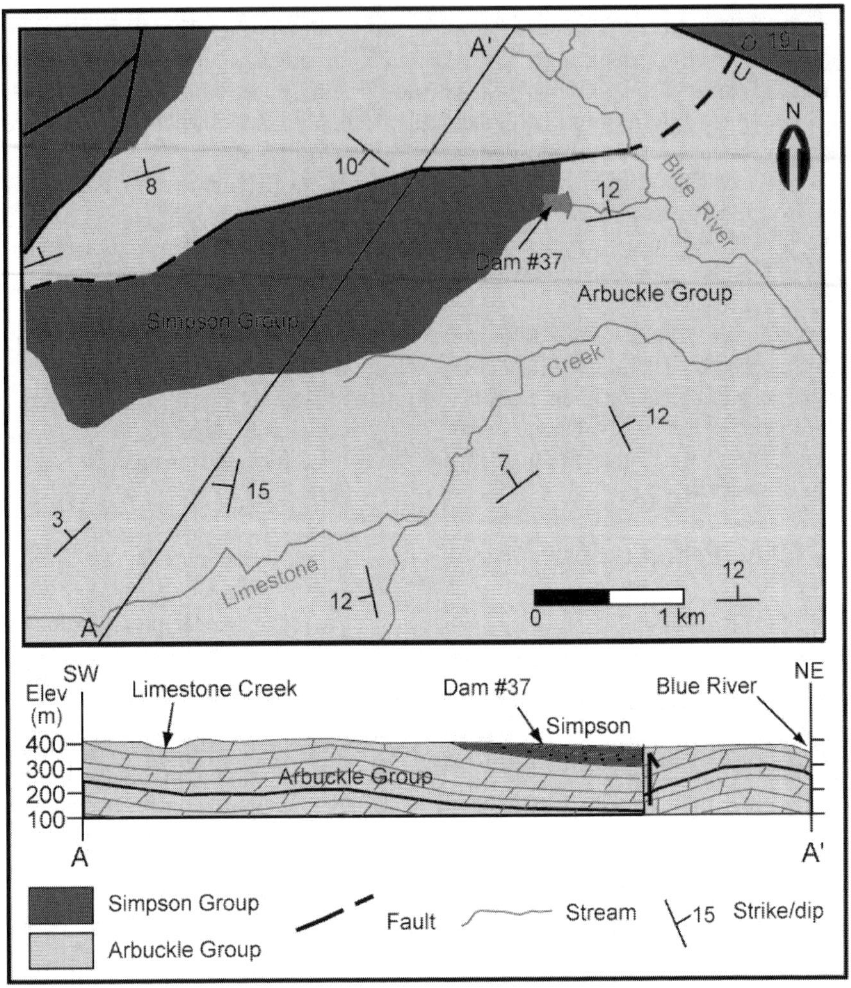

Figure 2. Geological map of the study area [adapted from 6].

This unit is thick in the north (adjacent to the fault) and thins to the south. The thickness of the Simpson Group changes from 1 m to 76 m in the study area [6].

Bogard [4] has identified three types of soils in the study area: the Port and Cleora soils (Pr), the Fitzhugh fine sandy loam (FhC2), and the Durant and Bates soils (DvC3). These low permeable soils are referred to as "vegetative soils" [4]. The thickness of the vegetative soil varies between 2 cm and 3.7 m.

The Arbuckle-Simpson aquifer is generally an unconfined carbonate aquifer (Arbuckle Group) with a sandstone (Simpson Group) overlying the Arbuckle Group in some locations. This fractured aquifer has good storage capacity in the soil zone and in the fractures in the aquifer and high transmissivity.

Due to heavy faulting in the study area, the occurrence and movement of groundwater in the Arbuckle-Simpson are strongly controlled by geologic structure. The association of springs with faults and other fractures indicates the significance of structural control on groundwater flow. The streams, like the Blue River and its tributaries, are fed by springs that issue from the aquifer [6].

The potentiometric level ranges from 335 m to 358 m [6] in the study area. The wells in the area show that groundwater is relatively shallow with estimated depths ranging from 14 to 43 meters.

4 Proposed artificial recharge system

Decision making protocols are already in place for the design of artificial recharge structures [7], but they are influenced by the presence of the existing upstream flood control dam. The reservoir can act as a settling basin and allows photodegradation of runoff water. With the dam in place, an "engineered sinkhole" is proposed to allow recharge, while filtering and providing a system that will require as little long term maintenance as possible, fig. 3.

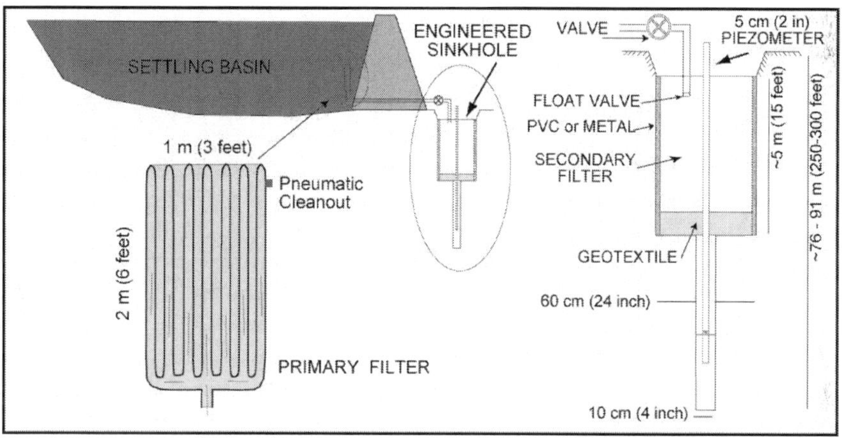

Figure 3. Schematic of the proposed artificial recharge system for Wingard Ranch.

The dam already in place to store fresh water from high precipitation events is the first component of the artificial recharge system. Besides this earth embankment dam, the recharge system has two additional components- the primary filter and the engineered sinkhole.

The primary filter is located in the basin next to the dam to filter fine sediments before the stored water will be diverted to the engineering sinkhole by a pipeline. The size of this filter is 1 meter by 2 meters. The main advantages of the primary filter are that the filter will remove sediment from the storm water when the dam is full and the filter will be cleaned by wind removing built up sediment when the dam is empty limiting maintenance. A pneumatic cleanout is also part of the design for maintenance if the filter does clog. The infiltrated water will then be transmitted via pipe to the engineered sinkhole. This pipe has a 10 cm diameter, a shut off valve, and a float valve to control the flow of the storm water.

The engineered sinkhole is composed of two parts- the secondary filter (upper) and the well (lower). The size of the filtration system is 60 cm diameter x 5 m long. The filtration system includes PVC or metal casing on the walls to block leakage, sand as a secondary filtration, and a layer of geotextiles at the bottom to prevent the sand filter from migrating into the well. The recharge well will be 20.3 cm (4 in) in diameter and approximately 76-91 m (250-300 ft) deep. A piezometer will be used to monitor groundwater levels and recharge water quality.

Although the proposed recharge well requires drilling and special design it is advantageous as the climate is sub-arid and land is costly. A recharge well will deliver surface waters directly into the deep aquifer and therefore bypass the low permeable soils and the rocks of the Simpson group. The main disadvantage that may exist is clogging by fines or biofouling. However, the infiltration system design helps to prolong the life by using the wetting and drying phases to limit these problems.

5 Discussion

Thousands of upstream flood control structures exist in the United States with 2101 in Oklahoma alone. These make excellent sites for recharge structures for several reasons: the water gathering structure that would be required for recharge is already constructed, the structure provides a pre-existing settling pond to reduce suspended solids from entering an aquifer, they would be expected to dry out and may not be allocated to significant surface uses, and since they dry out, they would limit the amount of growth of algae or other clogging substances allowing the structure to regenerate itself.

There are many options for recharge structures, however based on geologic and climate restraints, the majority of them are not a sound choice for the Wingard Ranch. The sub-arid climate, flat topography, high evapotranspiration, high clogging rates from microbial growth, potential impacts from cattle production, and low permeable soil covering are not ideal for spreading methods. This leaves the deeper methods such as wells.

As biological clogging will likely be the strongest maintenance issue for the recharge well, a modified design is provided. The design takes into account the benefits of using an upstream flood control structure. The primary benefit is the structure currently provided a settling basin, and is unlikely to flow continuously. Using this intermittent flow, the clogging issues can be minimized through careful design.

Cost is always an important aspect of choosing a system. The costs for evaluating, installing, operating, and maintaining a recharge structure on an upstream flood control structure include study costs, drilling costs for installing the system, filters and equipment to outfit the system, a weather station, sampling to monitor the system, and labor costs for permitting.

If the source water is suitable, a well will need to be drilled and tested to determine the intake rate of the formation at the site of the engineered sinkhole as the aquifer variability is significant in this type of formation. With the float valve in the system, the transmissivity can be quite low as the dam can remain full for several months while the water infiltrates the system. At this site the transmissivity is high enough to make the drainage time less than a week and perhaps less than one day for most rain events. The dam will also need to have a horizontal boring installed to allow the system to be installed. The total drilling costs are estimated at $66,000. The filter systems and valves to be installed are estimated at $3,000. An onsite weather station is estimated at $48,300. The remaining testing, construction, and permitting costs are estimated to be $22,000. Operation and maintenance costs for the structure include analysis of water samples and permitting, maintaining the weather station and transducers, and inspection and cleaning of the filter structures, if required. These costs are estimated at $23,000 per year, but may be reduced if the design works well and sampling and maintenance can be decreased after several years of operation.

There are two primary changes to the basin water budget that will be imposed on upstream flood control structure #37 if it is outfitted with an engineered sinkhole. First, runoff would be reduced to zero except in potentially extreme floods where the overflow to the dam would cause water to leave the system. Secondly, the evaporation and transpiration of the system would be reduced substantially. Impacts to the Blue River's environment should be minimal as the stream is ephemeral along this reach. A decreased amount of surface water input is unlikely to have a bearing on the Blue River, especially since only a nominal amount of the water from the conservation pool usually reaches it.

For the 287,000 m^3 of runoff entering the system, the distribution of the runoff was estimated as 5.6% (16,000 m^3) to direct evaporation, with an equivalent amount of transpiration near the lake. Natural groundwater recharge is assumed to be negligible, leaving the remaining 88.8% (255,000 m^3) of the budget as surface runoff out of the system. With the addition of the recharge system, this should reduce evaporation by approximately 90% to 1600 m^3 per year of the runoff budget. It is assumed that the recharge structure will do little to limit transpiration, so the budget for transpiration will remain unchanged for the runoff budget. If surface water outflow is converted to groundwater recharge, it will result in a recharge of 269,000 m^3 from the system. If water is traded in the basin, the system will act as a net gain to the water budget for the basin, and can potentially be used as a new resource in the basin for trading purposes.

6 Conclusions

Evaluation of the Wingard Ranch upstream flood control dam #37 indicates that the site will provide a suitable location for a recharge structure. The design of the "engineered sinkhole" structure is meant to simulate a sinkhole with better control on the filtering of the recharge water than exists for natural sinkholes in the area.

The site has constraints of high water quality in the aquifer. If the source water has lower quality, a recharge structure may not be possible, but the proposed settling and filtering mechanisms of the recharge design may solve any possible source water issues. As surface water already enters the aquifer through sinkholes, this is not expected to be a problem.

The residence time of recharge water is a second issue that needs to be addressed. If the residence time is too low, not only will the aquifer not be able to improve the water quality through common mechanisms, but the water won't be around long enough to make any significant differences in the basin as it will be discharged to the surface too quickly. The close proximity of Byrd's Mill Spring, which is used for municipal water supply, makes this evaluation more complex and a tracer test will be needed before the structure can be put online.

The water budgeting for the site indicates that approximately 269,000 m^3 of water could be recharged by the site, with 14,800 m^3 of water that would be lost to evaporation if no changes were made to the upstream flood control structure.

Acknowledgements

The authors wish to thank the late Harold Wingard for his support with this project. We are indebted to him for his help, cooperation and kindness. The authors also wish to thank the Scientific Research Projects Unit at Akdeniz University and the Bureau of Reclamation for financial support. Finally, we wish to thank Collins Balcombe and W. Robert Talbot for their invaluable assistance.

References

[1] Todd, D.K., & Mays, L.W., *Groundwater Hydrology*, John Wiley & Sons, Inc., New York, p. 251, 1959.
[2] Natural Resources Conservation Service, Upper Blue River watershed floodwater retarding structure site no. 37, Engineering Design Drawings, Stillwater, OK, 17 pp, 1999.
[3] Oklahoma Climatological Survey, http://climate.mesonet.org.
[4] Bogard, V.A., Soil Survey of Pontotoc County, Oklahoma, United States Department of Agriculture Soil Conservation Service: Washington D.C., 1973, pp. 1–34 , 1973.
[5] Terraserver, http://www.terraserver.com.
[6] Fairchild, R.W., Hanson, R.L., & Davis, R.E., Hydrology of the Arbuckle Mountains Area, South-Central Oklahoma, *Circular 91*, Oklahoma Geologic Survey, Norman, OK, 112 pp. & 2 plates, 1990.
[7] Gale, I.N., Neumann, I., Calow, R.C., & Moench, M., The Effectiveness of Artificial Recharge of Groundwater: a Review, *British Geologic Survey Commission Report*, Keyworth, Nottingham, 59 pp., 2002.

Evaluation of the Aardvark constant head soil permeameter to predict saturated hydraulic conductivity

E. Theron[1], P. A. L. Le Roux[2], M. Hensley[2] & L. Van Rensburg[2]
[1]*School of Civil Engineering and Built Environment,*
Central University of Technology, South Africa
[2]*Department of Soil Crop and Climate Science,*
University of the Free State, South Africa

Abstract

The saturated hydraulic conductivity (K_{sat}) is a key parameter for analyzing or modelling water flow and chemical transport in subsurface soil. Several methods were developed over the last thirty years to measure K_{sat} which considered unsaturated and/or saturated flow of water around the holes where measurements are made. Most methods make use of devices to measure the infiltration rate into the soil and in combination with numerical models predict the K_{sat} parameter. If maximum infiltration rate exceeds the infiltration capacity, runoff will be the consequence. Therefore maintaining a constant head means the rate of water supplied corresponds to the infiltration capacity. Commonly used devices for infiltration capacity measurements are infiltrometers, disc permeameters, sprinkler infiltrometers and different types of constant head permeameters. This paper evaluates the theory and practical application of the Aardvark constant head soil permeameter (ACHSP) in combination with the Glover equation for field measurement of K_{sat} above the water table. The Glover equation has been criticized during the eighties because only the saturated flow around the hole was considered in its development. Several papers published since 2002, however, demonstrated that the Glover equation results are relatively close to the results obtained by other models for most practical applications and therefore the use of the Glover equation is justified.

Keywords: saturated hydraulic conductivity, constant head permeameter, Glover equation, double ring infiltrometer.

doi:10.2495/SI100141

1 Introduction

The objective of this paper is to demonstrate that reliable saturated hydraulic conductivity (K_{sat}) values for subsurface soil horizons can be predicted by using the Aarvark constant head soil permeameter (ACHSP) in combination with the Glover equation. The field measurements were done at three selected sites at a depth of 800 mm and the results compared with values obtained by the double ring infiltrometer at the same depth for each of the three sites. The double ring infiltrometer is a more conventional method for determination of K_{sat}.

Hydraulic conductivity (K) is one of the more often used properties for evaluating soil suitability for different uses and for predicting the fate of anthropogenic materials applied on or in soil. It is therefore a key parameter for analysing or modelling water flow and chemical transport in subsurface soil (Mohanty *et al.* [1]). This parameter varies both spatially and temporally and is easily altered by management activities. Under saturated conditions the parameter is noted as K_{sat} and is assumed to be a constant for a given time and space within the soil (Amoozegar and Wilson [2]). Under both saturated and unsaturated conditions, the measure of the ability of soil to transmit water is described by Darcy's Law (Whitlow [3]).

The constant head permeameter method is the most versatile procedure for measuring K_{sat} of the unsaturated sone from near the surface to a few meters deep (Amoozegar and Wilson [2]). This technique is based on the direct application of Darcy's Law and the steady-state flow rate of water (Q) under constant head (H) at the bottom of a cylindrical hole of radius (r) is measured. In the original procedure this method took a few days and required a considerable amount of equipment and large quantities of water. However, due to development of equipment (e.g. ACHSP), development in the theoretical evaluation of subsurface flow of water and modifications of the field procedure in the last twenty years, steady-state flow rate from a small diameter cylindrical hole under a constant depth of water can now be reached with a few litres of water within two to three hours (Amoozegar [4]). Using a device such as the ACHSP the equation to calculate K_{sat} can be written in the form:

$$K_{sat} = CQ \tag{1}$$

The constant (C) must be calculated using a model. Amoozegar [4] proved that the Glover equation, as developed by R.E. Glover, is an appropriate model for determining K_{sat}. The Glover equation is used frequently to determine the K_{sat} of the unsaturated sone in combination with the constant head permeameter technique. The reason being, it only uses field measured data, is relatively simple and does not require estimation of any soil parameter. Provided that the values for Q, H and r are measured accurately, the ratio of $H/r \geq 5$ and the distance between the bottom of the hole and any impermeable layer below the hole (s) is $\geq 2H$, the Glover equation can be used to accurately calculate the K_{sat} value of subsurface soil (Amoozegar [4]). The constant (C) as given by the Glover equation is as indicated on the next page.

$$C = [\sinh^{-1}(H/r) - (1 + r^2/H^2)^{1/2} + (r/H)]/(2\pi H^2) \tag{2}$$

2 Field data collection

Three areas (2 x 2 m^2) with identical soil forms, similar soil forming conditions and homogeneous soil properties were identified as study sites. The subsoil horizon for which the K_{sat} values will be determined, has a deep red colour with minimum or no structure (apedal), is well drained and provides excellent to moderate cultivation opportunities. The location of the sites is within 1200 ha of land which was surveyed to delineate land use types for irrigation. This land formed part of a 3000 ha project to evaluate land suitable for irrigation in the Free State and adjacent Northern Cape Provinces in South Africa. In the mentioned project, the double ring infiltrometer was used to determine the K_{sat} values at selected sites. The study sites where K_{sat} values will be determined using the ACHSP are the same as three of the sites selected for the 3000 ha project.

The flow rate of the water into the subsoil horizon, occurring in the region of 0.2 m up to 1 m below the diagnostic topsoil horizon, was measured. The areas were levelled and a hole of radius (r) 3.5 cm was dug to the desired depth of 800 mm using a hand auger. After cleaning the bottom of the hole, the installation of the ACHSP was done and a constant rate of water flow into the soil under a constant depth/head of water was established. As depicted in figure 1, the constant head of water at the bottom of the hole is set at the desired level by moving the tube up or down with a cord. To ensure that H/r is ≥ 5 with a fixed auger radius of 3.5 cm, it was necessary to predetermine the distance between the bottom of the tube and the bottom of the hole (h_2) in the laboratory. The minimum constant head must be equal or greater than 17.5 cm to meet the above-mentioned requirement. Therefore it was decided to use a height of 12 cm (h_2) to ensure a corresponding value of 25 cm for H. The depth of water in the hole was measured at least twice to assure that a constant head is maintained throughout measurement. A minimum of 20 L and maximum of 40 L was needed to reach a steady-state flow rate of water within one to three hours. The fact that the soil is well drained influenced the amount of water needed and the time to reach steady state of water flow rate. The flow of water into the soil was determined by a pressure meter water gauge (probe) which recorded the height difference (h_1) of the water in the reservoir every minute. The conversion factor of height to volume of water (L) was determined in the laboratory as 0.013. The flow rate was now calculated by dividing the volume by one minute and then converted it to a flow rate of centimeter^3/ hour (see table 1). The position of the auger holes at the study sites are indicated in figure 2. The site was a distance of 3 m from the area where the double ring infiltrometer measurements were taken as part of the 3000 ha project. Four measurements at a depth of 800 mm were taken (P1 – P4) to correspond with the depth of the calculated values measured with the double ring infiltrometer. The auger holes were positioned as such that the minimum distance between any of the positions never exceeded 1.5 m. The reason was to minimise/prevent any effect the soil water regime around the different cylindrical holes may have on each other after steady state of water flow was reached.

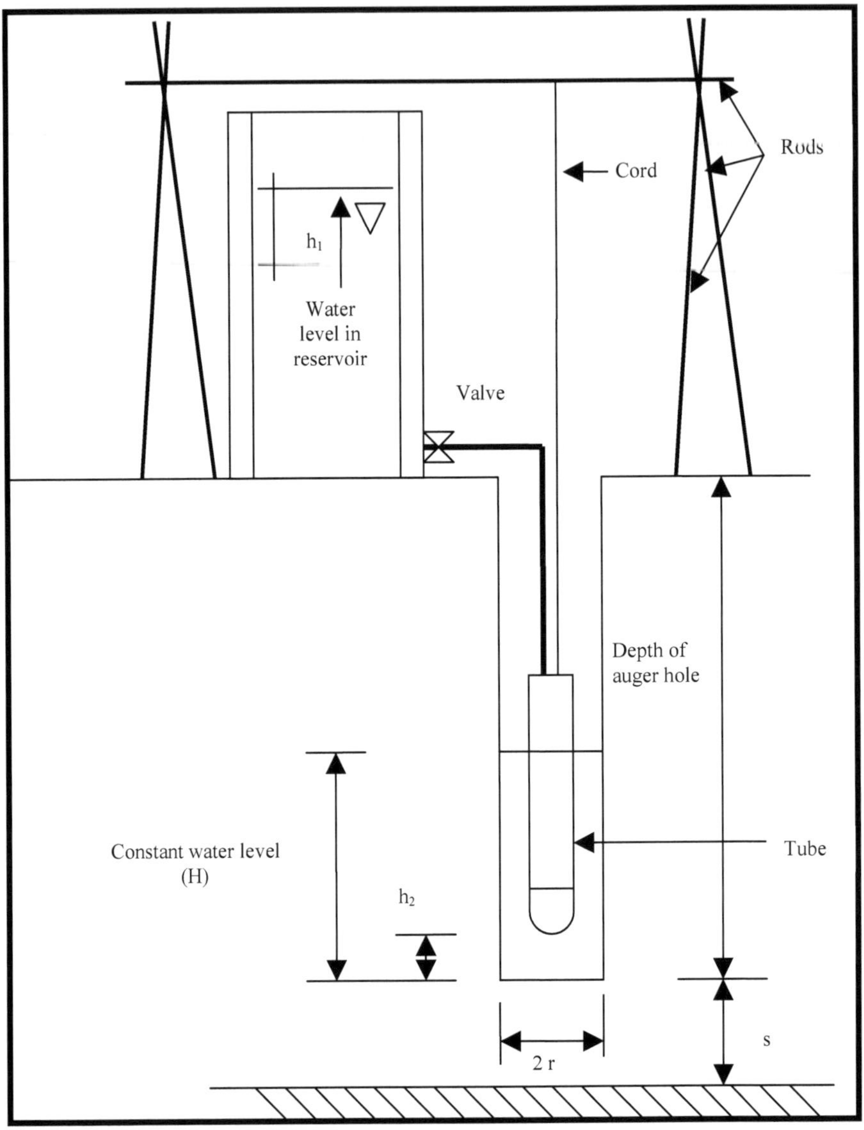

Figure 1: Schematic diagram of the Aardvark constant head permeameter.

3 Results and discussion

From the three sites under investigation only the dataset of site 3 could be used. This was due mainly to malfunction of the probe (pressure meter water gauge) and corrupt data. Valuable lessons, however, were learned which will come to

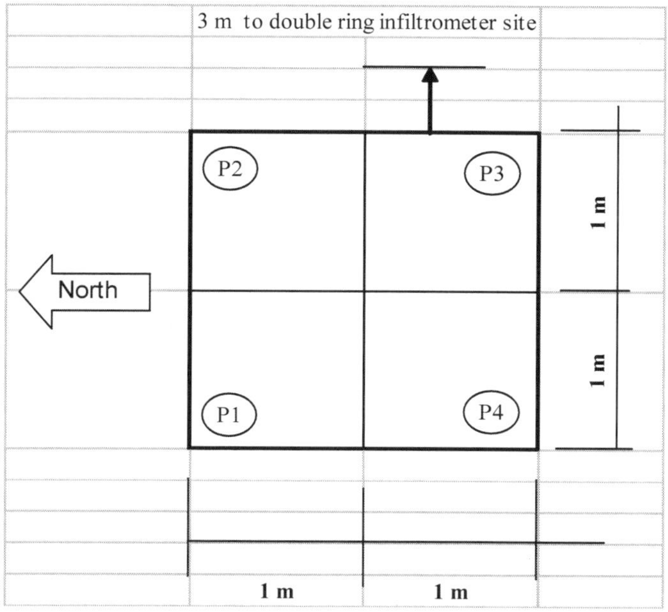

Figure 2: Position of auger holes.

good use in the follow up studies. The dataset of position one (P1) at site 3 for the first 26 minutes is given in table 1. The graph (figure 3) indicates flow rate versus time and an approximation of when steady state of water flow rate is reached. For P2 – P4 only the graphs are illustrated in respectively figures 4, 5 and 6.

In order to calculate K_{sat} by using eqn. (1), it is necessary to first calculate C as given by the Glover equation and then determine Q. The values necessary to calculate C are the constant head and the radius. As discussed in the section on field data collection the values are 25 cm and 3.5 cm respectively. The calculation of C by eqn. (2) is:

$$C = [\sinh^{-1}(25/3.5) - (1 + 3.5^2/25^2)^{1/2} + (3.5/25)]/(2\pi 25^2)$$
$$= 0.00046 \text{ cm}^{-2}$$

Logarithmic trendlines were added to the graphs to determine the reliability of the data which was used to determine Q. The type of data determines the type of trendline which should be use. A logarithmic trendline is a best-fit curve line when the rate of change in the data increases or decreases quickly and then levels out. A trendline is most reliable when its R^2 value is at or near 1. This value shows how closely the estimated values for the trendline correspond to the actual data. The R^2 value is displayed on the graphs. The values of Q for P1 – P4 as determined from the graphs ("constant" trendline value) are given in table 2. K_{sat} can now be calculated for each of the four positions with eqn. (1). These values are also indicated in table 2.

Table 1: Dataset (26 min) of site 3.

Time	Date	Pressure mWG	Height difference (mm)	Vol (L)	Time (h)	Q (cm^3/h)
					CF =	*0.013*
12:55:33	12.03.2009	0.289	0.012	0.92	0.02	55384.61538
12:56:33	12.03.2009	0.278	0.011	0.85	0.03	50769.23077
12:57:33	12.03.2009	0.268	0.010	0.77	0.05	46153.84615
12:58:33	12.03.2009	0.258	0.010	0.77	0.07	46153.84615
12:59:33	12.03.2009	0.247	0.011	0.85	0.08	50769.23077
13:00:33	12.03.2009	0.238	0.009	0.69	0.10	41538.46154
13:01:33	12.03.2009	0.229	0.009	0.69	0.12	41538.46154
13:02:33	12.03.2009	0.220	0.009	0.69	0.13	41538.46154
13:03:33	12.03.2009	0.210	0.010	0.77	0.15	46153.84615
13:04:33	12.03.2009	0.201	0.009	0.69	0.17	41538.46154
13:05:33	12.03.2009	0.192	0.009	0.69	0.18	41538.46154
13:06:33	12.03.2009	0.184	0.008	0.62	0.20	36923.07692
13:07:33	12.03.2009	0.175	0.009	0.69	0.22	41538.46154
13:08:33	12.03.2009	0.166	0.009	0.69	0.23	41538.46154
13:09:33	12.03.2009	0.157	0.009	0.69	0.25	41538.46154
13:10:33	12.03.2009	0.149	0.008	0.62	0.27	36923.07692
13:11:33	12.03.2009	0.14	0.009	0.69	0.28	41538.46154
13:12:33	12.03.2009	0.132	0.008	0.62	0.30	36923.07692
13:13:33	12.03.2009	0.123	0.009	0.69	0.32	41538.46154
13:14:33	12.03.2009	0.115	0.008	0.62	0.33	36923.07692
13:15:33	12.03.2009	0.108	0.007	0.54	0.35	32307.69231
13:16:33	12.03.2009	0.099	0.009	0.69	0.37	41538.46154
13:17:33	12.03.2009	0.091	0.008	0.62	0.38	36923.07692
13:18:33	12.03.2009	0.084	0.007	0.54	0.40	32307.69231
13:19:33	12.03.2009	0.076	0.008	0.62	0.42	36923.07692
13:20:33	12.03.2009	0.068	0.008	0.62	0.43	36923.07692
13:21:33	12.03.2009	0.061	0.007	0.54	0.45	32307.69231

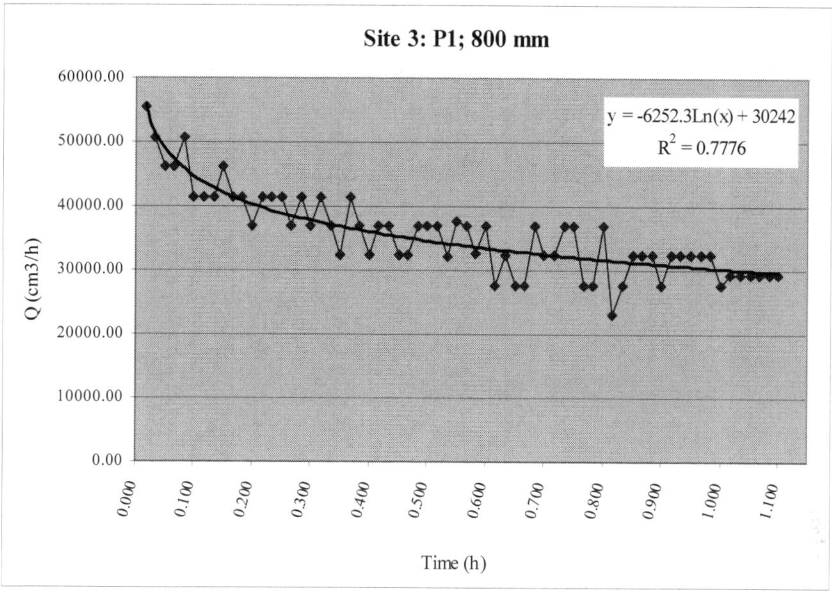

Figure 3: Flow rate measurements at position 1, site 3.

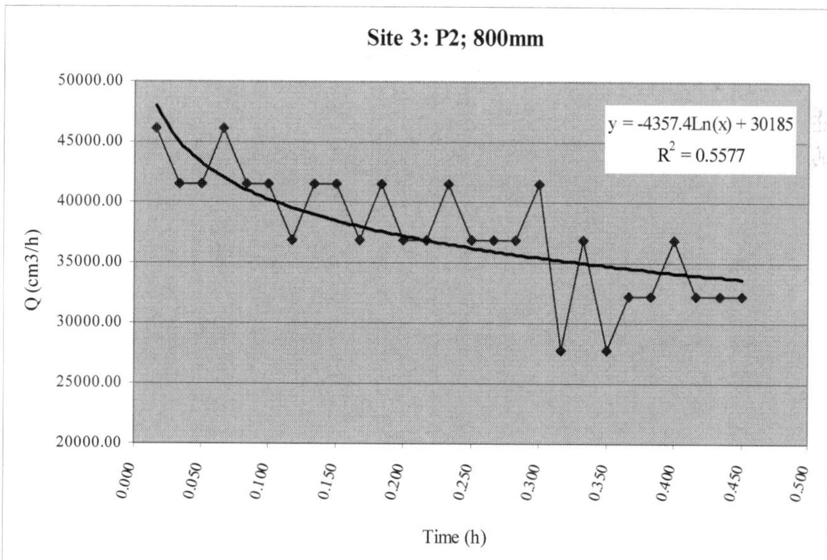

Figure 4: Flow rate measurements at position 2, site 3.

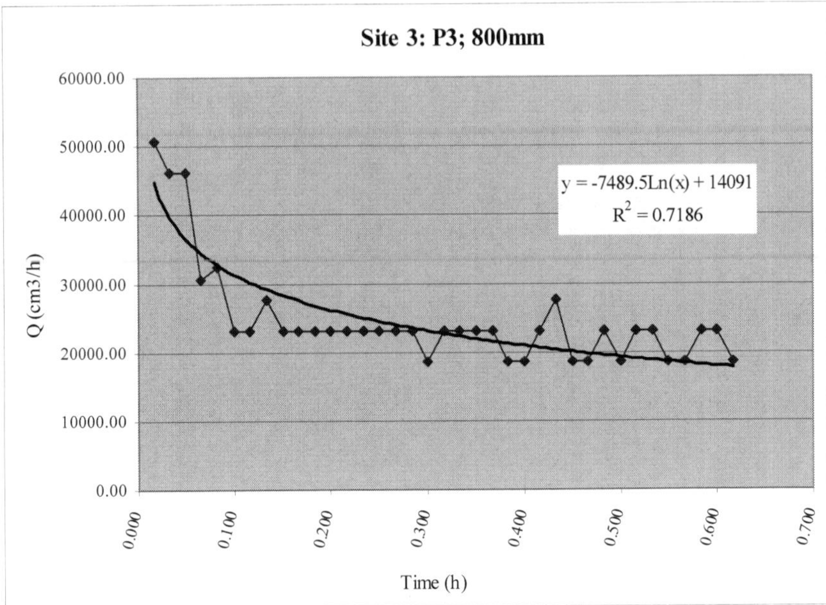

Figure 5: Flow rate measurements at position 3, site 3.

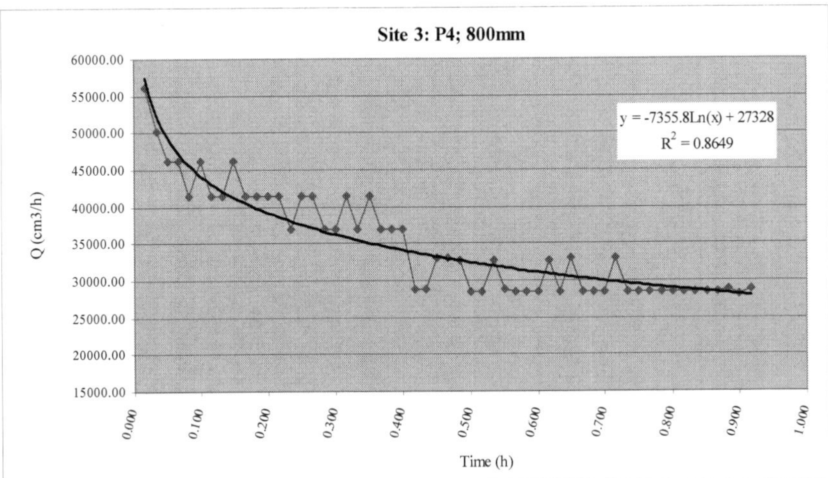

Figure 6: Flow rate measurements at position 4, site 3.

The objective of this paper was to demonstrate that reliable (K_{sat}) values for subsurface soil horizons can be predicted by using the ACHSP in combination with the Glover equation. The field measurements as done at site 3 at a depth of 800 mm can now be compared with values obtained by the double ring infiltrometer at the same depth, for this specific site. In the section on field data collection it was stated that site 3 is within 1200 ha of land which formed part of a 3000 ha project to evaluate land suitable for irrigation in certain provinces in South Africa. In this project, the double ring infiltrometer was used to determine K_{sat} values. Table 3 gives the values obtained for K_{sat} in this project at site 3.

Table 2: Steady state flow rate and saturated hydraulic conductivity at site 3.

Auger Position	Q (cm^3/h)	K_{sat} (cm/h)	K_{sat} (cm/day)
P1	30 000.0	13.7	329.0
P2	32 500.0	14.9	356.4
P3	18 460.0	8.4	202.4
P4	28 400.0	13.0	311.4

Table 3: K_{sat} values at site 3 as obtained by the double ring infiltrometer.

Rep1					
min	cm	mm/h	cm/d	cf	cm/d
0.96	1	625.00	1500.00	0.0006667	1428.7
1.01	1	594.06	1425.74	0.0007014	1358.0
1.11	1	540.54	1297.30	0.0007708	1235.6
1.73	1	346.82	832.37	0.0012014	792.8
1.92	1	312.50	750.00	0.0013333	714.3
2.19	1	273.97	657.53	0.0015208	626.3
1.97	1	304.57	730.96	0.0013681	696.2
2.03	1	295.57	709.36	0.0014097	675.6
2.22	1	270.27	648.65	0.0015417	617.8
0.35	1	1714.29	4114.29	0.0002431	3918.7
0.9	1	666.67	1600.00	0.0006250	1523.9
0.95	1	631.58	1515.79	0.0006597	1443.7
1.34	1	447.76	1074.63	0.0009306	1023.5
2.61	1	229.89	551.72	0.0018125	525.5
2.15	1	279.07	669.77	0.0014931	637.9
2.71	1	221.40	531.37	0.0018819	506.1
2.16	1	277.78	666.67	0.0015000	635.0
2.96	1	202.70	486.49	0.0020556	463.4
3.15	1	190.48	457.14	0.0021875	435.4
2.73	1	219.78	527.47	0.0018958	502.4

The average value determined in the 3000 ha project for K_{sat} was given as 467.1 cm/d.

4 Conclusion

A vast number of field methods have been developed over the past thirty years to measure the saturated hydraulic conductivity of soils above the water table. These methods, which include the constant head permeameter, air-entry permeameter, intake rate infiltrometer, double ring infiltrometer, velocity permeameter or falling head permeameter as well as several others, have been used with varying degrees of success. The most significant practical limitations of all these methods usually include an installation time of up to several hours, a measurement period of up to several days and water requirements of up to a 1000 L or more per measurement. In addition, considerable equipment and usually more than one operator are required to conduct tests. The development of the ACHSP has removed most of these limitations. The apparatus is lightweight and simple to install, it is inexpensive to construct and can be operated by one person. All the necessary field data can be collected within 1 to 2 hours per measurement and smaller quantities of water are required.

In the ACHSP technique, the saturated hydraulic conductivity is calculated after determining the steady-state rate of water flow from a cylindrical hole of known radius under a constant depth (head) of water at the bottom of the hole. The resulting outcome of the K_{sat} values as determined by the ACHSP is fairly close to the K_{sat} values as calculated by the more conventional double ring infiltrometer (356 cm/d compared to 467 cm/d). It seems if the ACHSP underestimate the K_{sat} values relatively to the double ring infiltrometer, however, it may be concluded that due to all the practical advantages and closely correlated values, the ACHSP is a reliable device to be used in field measurements to determine reliable saturated hydraulic conductivity values.

References

[1] Mohanty, B.P., Shaggs, T.H. & Van Genuchten, M.Th. Impact of saturated hydraulic conductivity on the prediction of tile flow. *Soil Sci Soc. Am,* **J. 62**, pp. 1522–1529, 1998.

[2] Amoozegar, A. & Wilson, G.V. Methods for measuring hydraulic conductivity and drainage porosity. *Soil Sci Soc. Am,* pp. 1149–1205, 1999.

[3] Whitlow, R. *Basic soil mechanics,* Henry Ling Education Limited, Great Britian, 2001.

[4] Amoozegar, A. Soil Science Society of North Carolina proceedings, **Vol XLV**, 2002.

Section 3
Sustainable irrigation – farmers' actions, environmental benefits and socio-economic consequences

Special session chaired by H. Bjornlund

Sustainable irrigation – a new challenge?

V. Bjornlund[1] & H. Bjornlund[1,2]
[1]*Department of Economics, University of Lethbridge, Canada*
[2]*University of South Australia, Australia*

Abstract

Irrigation has been practiced for some six or seven thousand years in various parts of the world. When sedentary farming emerged in areas along major rivers in the Fertile Crescent, the Indus Valley and China, irrigation soon followed to sustain food production through dry periods. Many of the areas suitable for irrigation were prone to flooding and therefore canals and other structures were not only needed to supply water for irrigation, but also to drain excess water from the fields, and for flood protection. Such structures were well beyond individual farmers and more complex societies emerged. The term 'sustainability' was not known to the first irrigators; they experimented, observed, learned, changed and adapted to their physical environment and changing climate. In some instances, the physical conditions were not suitable for intensive irrigation. Hence, for a number of reasons such as deforestation, soil erosion, salinization and siltation, some hydraulic societies ceased to exist, while in other instances continued for millennia until irrigation intensified due to increased population pressure resulting in changing technologies and practices. In some places irrigation has continued in the same location until today. This paper explores these experiences from around the world and tries to draw lessons for today's irrigation communities. We find that: decentralizing is better than centralizing; diversity is better than uniformity; and local self-ingenuity and self-reliance yield superior results. If exogenous technologies or knowledge are applied, it should be in conjunction with, and adapted to, local conditions, knowledge and cultures. We also find that many of the mistakes irrigators made thousands of years ago are being repeated today and, consequently, we are suffering the same effects – there seems to be a breakdown in our historical memory.

Keywords: sustainable irrigation, Mesopotamia, Egypt, Oman, India.

WIT Transactions on Ecology and the Environment, Vol 134, © 2010 WIT Press
www.witpress.com, ISSN 1743-3541 (on-line)
doi:10.2495/SI100151

1 Introduction

The concept of sustainable development gained prominence in 1987 with the Bruntland report *Our Common Future* [1], which refers to sustainable development as development that meets the needs of the present without compromising the ability of future generations to meet their needs. However, the problem of meeting the current generation's need for food and fresh water has been a challenge for mankind since the beginning.

Sedentary farming developed along the world's major rivers in Mesopotamia, Egypt, the Indus Valley, China, and in Mesoamerica. The new farmers soon learned that production could be increased if water was made available when natural precipitation was inadequate [3]. Under these circumstances, farmers could produce more food than they needed for subsistence. This made it possible to release human resources from food production and enabled the emergence of full time trades people and managers who could spend their time securing food production through the construction of more complex, large-scale irrigation structures. This required more complex and organized societies that developed division of labour, laws, and political hierarchies. The development of larger scale irrigation therefore coincided with the first urban societies, termed hydraulic societies [4]. Among them are the Sumerian in Mesopotamia, the Egyptian, and the Harappan Civilization in the Indus Valley. The first written law, the law of Hammurabi from 1,784 BC, includes 300 sections concerning irrigation [5].

Since the first hydraulic societies, history has seen the emergence of many civilizations based on irrigation which initially prospered. However, irrigation has problems which can cause the downfall of societies that rely on it. The construction of irrigation infrastructure and the introduction of intensive irrigation have caused sedimentation of canals, water-logging, and salinization and created havoc in places such as Mesopotamia. In some instances, salinity and water logging resulted in declining yields in the short term, while in others it took centuries and even millennia. In most cases, population pressures and/or confrontation with exogenous technologies resulted in the adoption of irrigation practices which were unsuitable for the local conditions.

Historical records make it clear that the destructive impact of man was already apparent five thousand years ago in China and the Middle East, when the first period of accelerated, man-made erosion began [6]. Cutting or forests, cultivation of sloping lands, overgrazing and salt accumulation were responsible for most of the land degradation in the uplands of China and around the Fertile Crescent. Already 1,600 BC, the Chinese Emperor You acknowledged this when he said *to protect your rivers, protect your mountains*.

2 Examples of early irrigation systems/societies

2.1 Mesopotamia

Irrigated agriculture in Mesopotamia has been practiced along the banks of the Euphrates and Tigris Rivers as far back as the 4th or 5th millennium BC [7].

Some 10,000 years ago, sedentary farming started in areas with sufficient rainfall. However, around 4000 BC local people had moved to the dryer plains of Lower Mesopotamia. They found that crops would not grow with the natural precipitation alone. Initially, their response was to sow seeds following the spring floods to benefit from the moisture in the ground. However, in some years dryness caused the crops to die before harvest.

In response to this problem they built diversion canals to convey water from the Euphrates and Tigris to their fields, allowing them to apply water when needed and also to expand production onto areas not covered by the spring floods. They practiced irrigated basin agriculture which included alternate-year weed-fallowing during which the land is temporarily abandoned, with the result that wild plants take over and dry out the subsoil to a depth of 2 meters, thus lowering the water table and preventing the salt from coming to the surface [8]. These irrigation networks consisted of short canals, keeping the irrigation in close proximity to the main watercourse. One of the benefits of the short canals was that siltation was not a big problem and could easily be cleaned up. The small and simple systems meant that canals could be constructed, maintained, operated, and replaced by the local group of farmers without the need for a powerful centralized government. The ability to control the two worst problems associated with irrigation, siltation and salinization, meant that this system was in operation over a very long period of time [6].

To satisfy the need for water and food for an increasing population and the expanding urban societies such as Ur, there was a massive and unprecedented increase in building irrigation works in the Euphrates Valley sometime during the 3rd millennium BC. Vast canals and complex systems of branch canals were constructed to supply water to unused desert land and depression areas.

A particularly devastating impact was experienced following the building of the 300 km long Nahrwan Canal around 2400 BC in Southern Iraq. The system consisted of a complex network of canals skilfully planned and regulated. The methods employed appear to mainly involve gravity flow from the Euphrates to irrigate land previously supplied from the Tigris. The canal was so large that it was referred to as 'the Tigris' [9]. These massive systems included thousands of brick sluice gates along its branches. In short, it was a system that could only be built and maintained by the resources of a specialized, centralized and powerful government. The result was a total shift in the concept of irrigation away from the previously small and locally controlled systems toward a wholesale reshaping of the physical environment [6]. This resulted in seepage, flooding and over-irrigation. Combined with the abandonment of fallowing and insufficient drainage, the groundwater level started to rise, eventually causing extensive problems with soil salinity as is evidenced in temple records of the time [6]. The long canals also resulted in widespread siltation of canals and the deposit of infertile silt on the land, further reducing the fertility of the soil.

As a consequence, there was a gradual but marked reduction in the production of salt-sensitive wheat, the preferred food crop, and an increased production of the more salt-tolerant, but less attractive barley. Around 3500 BC, archaeological evidence suggests that an equal amount of wheat and barley was grown in

Southern Iraq. By 2500 BC wheat accounted for only one sixth of production, by about 2100 BC for no more than two percent, and by 1700 BC no wheat was grown at all. Not only did they suffer a shift from wheat to barley, but soil fertility also declined dramatically. Around 2400 BC, yields were about 2,537 kg per hectare; by 2100 BC it had declined to 1,460 kg, and by 1700 BC to 897 kg per hectare. Inevitably, the large-scale, centralized irrigation systems broke down under the strain of social upheaval [6] and a civilization disappeared.

The Babylonian King Hammurabi in 1784 BC introduced the first written law setting out the rights and obligations of all men, known as the 'Code of Hammurabi'. Reflecting the importance of irrigation the Code contained about 300 sections dealing with various aspects of irrigation. One of the laws said: "if a man is negligent in strengthening the banks of his field and has not maintained his banks and then a breach has occurred in his bank and so he has let the waters carry away (the soil on) the water-land, the man in whose bank the breach has occurred shall replace the corn which he has (caused to be) lost" [5].

The destruction of irrigated farmland in what today is known as Iraq is not isolated to ancient times. After the collapse of the ancient societies, irrigation continued on a smaller scale with many areas being abandoned as the old system deteriorated and was not maintained. Under The British Mandate (1914–1932) a slow revival and restoration of ancient works took place. Canals were dredged, regulators and weirs were constructed and the system was steadily extended and improved. At the same time, the *shaduf,* or the old fashioned waterwheels, was replaced with hundreds of diesel-pumps, and the land under perennial irrigation along both the Euphrates and Tigris were expanded. After The Second World War, an Irrigation Development Commission was formed to usher in a new era during which rapidly increasing oil revenues were used to invest in development of new irrigation schemes, dams and hydro-power, and reservoirs and flood control measures [10].

In more recent times, the upstream states of Syria and Turkey have dammed the Euphrates and Tigris to develop irrigation in their own countries. This has significantly reduced the volume of water flowing into Iraq and thereby also reduced water quality, especially resulting in increased salinity levels [11]. These new initiatives are causing further water logging and salinity affecting 8.5 million ha or 64% of the total arable land in Iraq, while 20-30% of irrigated land has been abandoned due to salinization [12]

Goldsmith concludes that "when (large-scale irrigation) regions are left in the uncontrolled possession of a landlord class, which is either of foreign origin or a partner in a precarious alliance with a foreign conqueror, rural investments are in danger of being neglected, because the landlords inevitably go for quick profits and liquid assets. In extreme cases the result is starvation and depopulation" [7].

It has often been argued that the collapse of the ancient civilization in Mesopotamia with their well-organized and productive farms was associated with a changing climate and drying of the land. However, while there is ample geological and archaeological evidence that this area was wetter and likely more fertile in the distant past, there is no conclusive evidence that these changes took place in historic time and certainly not over the last 10,000 years. Hence, the decay of a civilization and the disappearance of what was likely the oldest and possibly the most advanced hydraulic society must be found in other factors. It is most likely that it was associated with the actions that people took over a long period of time and in their inability to adopt the way they farmed as they experienced the negative consequences of their actions. There is no reason to believe that the farmers of the time did not understand cause and effect in their landscape, that when they extended the canals and abandoned fallowing they caused the problems which eventually resulted in their demise [13].

2.2 Egypt – the Nile

About the same time as in Mesopotamia, farmers started irrigating along the Nile, utilizing the flash spring floods to inundate their fields in the narrow and low-lying Nile Valley. It was mainly a natural phenomenon but farmers dug flat-bottomed basins along the river and a system of canals to flood the fields from the upstream ends of the basin and drain the water back into the river at the downstream end. The canals were also used to convey the water to fields further away from the river. However, this was limited to the low lying areas of fertile soil which could be reached without lifting devices. Irrigation was also limited to once a year to coincide with the natural flood when water came down from the mountains in Ethiopia and Uganda. Water was in abundance and there was no opportunity to irrigate outside this period, thus preventing continuous cropping.

Inundation irrigation worked sustainably for 5,000 years because of its simplicity, efficiency, and cost-effectiveness. The Nile performed three important functions; i) it provided irrigation water, ii) it disposed of drainage water; and iii) it rejuvenated the soil by depositing fertile silt from the upstream mountains. Further, large volumes of water were used only once a year which leached potential salt deposits out of the soil [14].

This system supplied Egypt with ample food up until the 18th century. The population and irrigated area in Egypt prior to 1800 peaked in about 100 AD when Egypt supplied the Roman Empire with vast quantities of grain [14]. In the 19th century Egypt experienced a population explosion, and an increase in the use of agricultural land for cotton production for export to England [15]. As a consequence, food production started to fall short of needs. To expand irrigation for food and cotton production a series of dams were built across the Nile north of Cairo, followed by a number of other dams. Dam building culminated when British engineers built the first Aswan Dam in 1904, which was subsequently expanded twice by 1934 and completed when the high Aswan Dam was constructed from 1964 to 1970.

These developments fundamentally changed irrigation and the flow of the River Nile. Annual inundation and cropping was replaced by continued irrigation with two or three crops a year. The Nile was controlled and did not flood anymore; instead, the fields were continuously wet and were never thoroughly leached to flush out the salt from the soil. In addition, the water no longer deposited the nutritional silt, as this was now trapped behind the dam wall. This resulted in decreasing fertility, a buildup of soil salinity and water logging [16]. Already in 1928 scholars warned about the threat of land salinization and that the big dams would hold back the silt that had replenished the fertility of the floodplain for millennia and the cotton harvest started to decline [17]. A crude estimate of the amount of crop yield reduction attributed to salinization and waterlogging in the Nile Valley indicates it to be at least 17 percent on 80 to 90 percent of the old irrigated land. In the newly irrigated land, the figure is likely to be a minimum of 25 percent [18]. As in Mesopotamia, it can be observed that human interactions altering the natural flow of the river caused soil degradation, most importantly in this instance water logging and salinization. Whereas increased siltation in Mesopotamia was part of the problem as the silt was unfertile, in Egypt the fact that the large dams prevented the fertile silt from reaching the fields was part of the problem. As in Mesopotamia, these actions were taken in response to increased population pressures but also as a result of a political decision made by a colonial power to increase export of cotton and unwillingness to adapt when the impacts started to appear.

2.3 The qanat system

For at least the last 2500 years underground water mobilization systems, known as qanat systems have been constructed and used throughout the Arab World.

Qanat technology originated in the highlands of western Iran, northern Iraq, and Eastern Turkey, possibly in connection with early mining ventures in that region. These are areas where water scarcity is extreme, but maybe more importantly, where the availability of fertile land was even scarcer and located in areas at some distance from the water source. Hence, water needed to be conveyed to where the fertile land was, thereby allowing people to inhabit areas with little rainfall. Although it originated in Iran, the system spread throughout the Middle East, the Arabic Peninsula, North Africa, and along the old Silk Route [19].

This discussion is based on qanat systems in Oman, where they are locally know as falaj systems. There are three types depending on the source of water: i) Daudi systems supplied by groundwater and conveyed by underground tunnels; ii) Ghaili systems supplied by open channels fed by surface water in oases; and iii) Aini systems supplied by springs in the foothills [20]. Common among the three systems is that they are based on gravity and water cannot be extracted faster than it is naturally replenished. Therefore, the systems are sustainable.

This paper concentrates on the Daudi system. The system extracts water from the top of aquifers high in the mountains and brings the water in underground tunnels down to the lower areas, and from there in open channels into villages and onto the small fields, mainly for irrigation of date palms, but also for other cash or subsistence crops [20, 21]. This system has several unique features for sustainable irrigation from environmental, social and economic perspectives.

The Daudi system cannot mine the aquifer as the off-take is located high in the aquifer profile: consequently, the system only harvests the overflow or the renewable yield from the aquifer. The system requires skill to establish and maintain, but it is entirely gravity fed so there is no pumping cost and potential disruption caused by shortage of power or spare parts. This reduces the monetary expense associated with running the system and provides a secure water supply.

In most systems only a small proportion of the water rights are privately owned. In one system, visited by one of the authors, only 15% of the water rights are privately owned, 10% are owned by the village community and the rest by semi-public charity organizations. The water owned by the village community is sold to the farmers at auction every week and the proceeds are used to finance the maintenance and administration of the system. The majority of the water owned by semi-public institutions is sold to the farmers at annual auctions and the proceeds are used to pay for schools, police, priests and the poor.

In a sense, this system acknowledges the fact that all wealth is derived from the availability of water and, hence, the water is sold to farmers to fund the maintenance of the system and the running of the community. This system resolves many of the problems facing irrigation systems in developing countries, such as: i) fee collection and therefore adequate funding of operation and maintenance resulting in inefficient systems with reduced productivity; ii) equitable distribution of the wealth generated by the water. In this system equity is ensured as most farmers have to buy the water each year and the revenue is used to pay communal expenses; iii) risk sharing associated with reduced supply during periods of scarcity. This is also an integral part of the system as farmers have the right to the flow from the system during a specified period of time; during periods of scarcity the flow simply yields less water. During periods of severe drought, the watering cycle is extended from, for example, every seventh day to every tenth day. Hence, the cost and pain of scarcity is shared. The weekly auctions can then be used to redistribute who gets access to the limited flow; and iv) water for domestic use. This is also supplied by the system; as the canals enter the village, there are access points where villagers can take water for drinking and household purposes; further downstream are places where households are allowed to wash their clothes, after which point the water is diverted into the fields.

This system has operated for thousands of years [20, 21]. In Oman more than three thousand such systems were once in operation. Today many have run dry while others are currently under threat. So, why is this sustainable system under threat? In most instances this is due to the introduction of modern technologies that are enabling users to mine the aquifers. Private farmers are sinking tube wells into the same aquifers from which the qanat systems are supplied. Once the water level in the aquifer drops below the off-take for the qanat system, the flow of water ceases and the system runs dry, stopping the supply of water for both domestic and farming purposes as well as the flow of revenue supporting the communities.

2.4 India – a diversity of systems

Archaeological evidence suggests that irrigation systems have existed on the Indian subcontinent for at least 5,000 years. Reflecting the wide variability of rainfall, temperature, altitude, topography and soil types; the subcontinent has seen a wide variety of early settlements devising their own unique systems of water harvesting adapted to local conditions [22].

For example, in southern parts of India tens of thousands of village tanks (dams) were constructed. This suited local conditions in the region which were dominated by scanty rainfall, steep slopes with fast run-off, and soils with low water retention. There was, therefore, a need to be able to retain and store the water when it was available and apply it when the soil needed it at a rate at which it could be absorbed by the soil [22]. In other parts of the country, communities have devised various other means of water harvesting from rivers, aquifers and rainwater runoff to provide water for drinking and agriculture.

Evidence suggests that these traditional systems were very efficient and productive. A British survey of 2,000 villages in the province of Tamil Nadu found that between 1762 and 1766 some villages produced up to 12 tons of rice per hectare, a level of productivity which today has only been achieved in the most successful areas, during the most productive era of the Green Revolution [23]. The same document also reported that the availability of food per household was five tons compared to today's three-quarter tons. Emphasizing the productivity and efficiency of farming and irrigation in India is a report in the Ecologist Magazine by a British agricultural expert. He wrote the report based on a tour of India that he conducted in the 19th century to see how he could best advise Indian farmers how to improve their agricultural practices. He concluded that he could learn more from the Indian farmers than they could learn from him, as they knew so much about soil composition and health, pest control and water management, crop breeding, and all other aspects of agriculture [24].

The traditional irrigation systems suffered during three periods. When the East India Company ruled India, their primary interest was to extract resources from existing infrastructure and, therefore, they failed to invest in improvements and neglected to maintain infrastructure such as irrigation systems. To make matters worse, transport infrastructure such as roads and railways were often constructed on embankments which interrupted the traditional contour-based irrigation canals and tanks. It also cut off large areas from supply of loam-bearing water from rivers, leaving the land unproductive and sterile [25].

During the early days of the British colonial rule there was some investment in irrigation such as weirs, dams and barrages mainly in the Punjab region to produce export crops such as cotton, opium, sugar cane and wheat which generated most revenue to the state [26]. On the other hand, the British showed no interest in the traditional small scale peasant irrigation systems of wells, dams, small canals and tanks which had fed the Indian population for thousands of years. Various actions of the British Colonial system were detrimental to the continued use of traditional irrigation systems: i) British tax policies heavily penalized peasants who sank their own wells at their own expense; ii) the generally high land tax level during the early colonial period left local communities with no financial ability to maintain their irrigation systems; iii) the taxation system destroyed the social fabric and system which had enabled local communities to construct and maintain irrigation system themselves. Indeed, British rule relieved local political chiefs from their obligations to invest in community resources and public institutions such as tank systems; iv) modern irrigation systems when constructed were in many instances deliberately situated in ways that supplanted other prevailing irrigation structures; and v) in some instances local farmers were actually forbidden to use their traditional system so as to ensure their payment to the new systems [27]. The refusal of the state to support local irrigation became a smouldering grievance everywhere in interior India. This disruption of the community structures was detrimental as traditionally irrigation systems in India have always been treated as communally managed common resources [28].

Since Independence the Central Indian Government has shown relatively little interest in traditional small village irrigation systems. The Government has been far more interested in promoting the construction of large dams and in enhancing control over the resources by taking over the village tanks. This has led to a further breakdown of the community control and management practices that were so vital for the sustainability of the system. A substantial number of tanks acquired by the state have fallen into disrepair because village communities have stopped contributing the voluntary labour for desilting and upkeep of the tanks and the state is not able to perform or pay for the functions earlier undertaken by the people. By doing this, the state has also shown that it believes it is better to construct new large dams rather than revive the efficient network of the village tanks across India [29–33].

The impact of the government's program was amplified by its commitment to the new infrastructure and the fact that it promoted its use through a set of subsidies which made it more viable for the farmers to use the new systems than maintain the old systems. This has happened in a number of ways: i) subsidies for electricity supplies made it cheap to pump water from dams via canals or from tube wells deep in the ground; ii) subsidies for fertilizer which made farmers less dependent on the use of the tank silt which was used in the past as production enhancing material; iii) subsidies for water from the major dams reduced the dependence on tanks for irrigation; and iv) subsidies for diesel pumps also reduced the dependence on water from tanks [34].

The introduction of the Green Revolution in the late 1960s following severe drought and hunger further intensified irrigation, especially from groundwater in Punjab. Not only was the area under irrigation expanded but new high responsive varieties were introduced requiring the application of more water per hectare of crop. Finally, rice, a very water demanding crop, was introduced in Punjab. Most of this irrigation was supplied by tube wells which soon lowered the aquifers to levels not accessible by the traditionally dug wells and, hence, forced these smaller farmers to abandon irrigation, invest in tube wells or sell to neighbouring larger farmers who could afford such investments. As a consequence of these changes, aquifer levels are dropping at high rates, challenging future food production. Water tables are rising in many areas with associated salinity and other soil degradation problems. Since the introduction of the Green Revolution in India, severe land degradation is reported to have affected 58 million hectares in 1990 (20% of land). Punjab and Hyryana are experiencing declining wheat yields, with soils becoming unsuitable for wheat cultivation [35].

3 Discussion and conclusions

Irrigation has been practiced for more than 6,000 years in various parts of the world to feed and clothe a growing population. In many places irrigation was initially sustainable for many years while farming practices were used which allowed the land to recover between crops and irrigation applications were scheduled so that salinity build up was prevented. Once irrigation intensified and continuous irrigation and cropping was introduced, the land did not have time to recover and salinity and water logging problems started to reduce productivity and in some instances resulted in the collapse of societies. In Mesopotamia this started relatively early, while in Egypt it did not start until last century.

In Oman the qanat system worked both economically, socially and environmentally sustainably up until recent times, when the introduction of modern technology in the form of tube wells has caused the closure of many systems as aquifer levels are drawn down below the off-take of the qanat system.

In India, British Colonial rule destroyed in a variety of ways many traditional irrigation systems which had successfully fed the local populations for many generations. Not only did they destroy or disrupt the physical structures, they also destroyed the social and community structures that had kept these systems in operation. The change in priority from feeding the local population to maximizing export production and revenue from land taxes were the major reasons for this development. Since Independence, the Indian Government has continued the destruction of the local systems through its emphasis on very large systems and associated subsidy systems favouring the new mega systems at the expense of the local system.

What we can learn from all of this is that we have to respect existing physical and social structures and not replace them with modern technology until we are sure that we fully understand how they operate and how the modern technology is going to interrupt the existing systems and structures. Pursuit of short-term economic gains does not justify causing environmental damage as such harm imposes social and economic costs in terms of higher health bills or diminishing agricultural returns on future generations [36]. The most successful systems rely on local knowledge and organization. We conclude that

- decentralizing is better than centralizing;
- that diversity is better than uniformity; and that
- self-ingenuity and self-reliance yield superior results.

If exogenous technologies or knowledge are applied it should be in conjunction with, and adapted to, local conditions, knowledge and cultures, drawing on generations of knowledge of the local conditions.

Maybe most importantly, we should appreciate that the problems discussed in this paper have been caused by human interference with natural systems and by an inability or unwillingness to take appropriate measures when negative impacts of this interference emerged. This problem is not isolated to early civilizations and developing countries. Irrigation experiences from the last couple of hundred years in California, Australia and elsewhere have repeated the same type of manipulation with natural systems as discussed here and have been experiencing the same consequences of water logging and salinization, causing millions of hectares to be taken out of production. These developments today, as they did then, represent the most serious challenge to future food security and the future of our societies as we know them.

Since these problems are caused by human actions and behaviour, only by changing this behaviour can we rectify the problems. The issue will ultimately be whether we have the political will and foresight to do it.

References

[1] WCED, World Commission on Environment and Development, *Our Common Future*. Oxford University Press, Oxford U.K, 1987.

[2] Diamond, J. *A Short History of Everybody for the Last 13.000 Years*. W. W. Norton & Company, New York, 1997.

[3] Hassan, F. A., The dynamics of a riverine civilization: a geoarchaeological perspective on the Nile Valley, Egypt. *World Archaeology*, 29(1), pp. 51-74, 1997.

[4] Wittfoget, K.A: *Oriental Despotism: A comparative Study of Total Power* Yale University Press, New Haven, 1957.

[5] El-Yussif, F. et al., Condensed History of Water Resources Developments in Mesopotamia. *Water International*, 8, pp. 19-22, 1983.

[6] Jacobsen, T. and Adams, R.M., *Salt and Silt in Ancient Mesopotamian Agriculture*. Science 128(334), pp. 1251-1258, 1958.

[7] Goldsmith, E and Hildyard, N.: *The Social and Environmental Effect of Large Dams*, ch 25, Wadebridge Ecological Centre, Cornwall U.K, 1984.

[8] Jacobsen, T., *Salinity and irrigation: Agriculture in Antiquity*. Report on Essential Results, Diyala Basin Agricultural Project, 1957-58. Malibu, Calif: Undena Publications 1982

[9] Adams R.M., A Synopsis of the Historical Demography and Ecology of the Diyala River Basin, Central Iraq. In Woodbury, R.B. Ed. *Civilizations in Desert Lands*, Anthrological Papers No 62, Department of Anthropology, University of Utah, 1958.

[10] J.H.G. Lebon: *The new irrigation era in Iraq*. Economic Geography 31 (1),

[11] Berna KIendiri, Belgin Cakmak and Yusuf Ucar: *Salinity in the Southeastern Anatolia Project (GAP), Turkey: Issues and Options*. Irrigation and Drainage 54: 115–122, 2005.

[12] Abul-Gasim, S. and M. Babiker: *Iraq's Food security: the sand dunes fixation project*. Desertification Control Bulletin, **33**, 2–10, 1998.

[13] K.H. Oedekoven: Forest history in the Near East. *Unasylva* 68(17), 1963.

[14] Postel, S., *Pillar of Sand: Can the Irrigation Miracle Last?* W.W. Norton & Company, New York, 1999.

[15] Wyn F. Owen: *Land and Water Use in the Egyptian High Dam Era*. Land Economics 40(3), pp 277-293, 1964.

[16] Hunt, R. C., Agricultural Ecology: The Impact of the Aswan High Dam Reconsidered, *Culture & Agriculture* 8(31), pp. 1 – 6, 1987.

[17] Carrier, E.H., *The Thirsty Earth*, Christophers, London, 1928.

[18] Dregne, H. E. 1986. Desertification of arid lands. In *Physics of desertification,* ed. F. El-Baz and M. H. A. Hassan. Dordrecht, The Netherlands: Martinus, Nijhoff.
[19] English, P.W., The origin and spread of "qanats" in the old World. *Proceedings of the American Philosophical Society*, 112, pp 170-181, 1968.
[20] Al-Marshudi, A.S.: The falaj irrigation system and water allocation markets in Northern Oman. *Agricultural Water Management* 91, pp.71-77, 2007.
[21] Zekri, S. and Al-Marshudi, A.S., A millenarian water rights system and water markets in Oman. *Water International*, 33(3), pp. 350-360, 2008.
[22] Aggarwal, A. and Narain, S., *Dying Wisdom: Rise, Fall and the Potential of India's Traditional Water Harvesting Systems*. A Citizen's Report No. 4, Centre for Science and Environment, New Delhi, 1997.
[23] India – Once Plentiful: Records reveal British schemes diminished crops and dismantled a native system of abundance. *Hinduism Today*, May, 1997.
[24] Sahtouris, E: Earthdance: Living Systems in Evolution, chapter 20: Sustainable Society. iUniverse 1999. Accessed 16th April 2009 at: http://www.ratical.org/LifeWeb/Erthdnce/chapter20.html.
[25] South Asian History, *History or the Indian Subcontinent: British rule and the legacy of colonization.* Accessed on 16th April 2009 at: http://india_resource.tripod.com/colonial.html .
[26] Davies, M., How famine was created. *Infochange*. Accessed on 16th April 2009 at http://infochangeindia.org/index2.php?option=com_content&do_pdf=1&id=5665.
[27] D'Souza, R., Water in British India: The Making of a 'Colonial Hydrology'. *History Compass 4/4*, pp. 621-628, 2006.
[28] Hardiman, D., Well Irrigation in Gujarat: Systems of Use, Hierarchies of Control, *Economic and Political Weekly*, 20 June 1998, p.1537.
[29] Shankari, U: Major Problems with minor irrigation, *Indian Sociology*, 25 (1), 85-111 1991.
[30] Sengupta, N: Managing common property: irrigation in India and the Philippines, Sage Publications, New Delhi 1991.
[31] Sengupta, N.: The Indigenous Irrigation Organisation in South Bihar. In Barah, B.C. Ed. *Traditional Water Harvesting Systems - An Ecological Economic Survey*, New Age International Publishers, New Delhi, 1996.
[32] *Singh, K.: Managing Common Pool Resources: Principles and Case Studies,* Oxford University Press, New Delhi. 1994.
[33] Von Oppen, M. and Subba Rao, K.V., *Tank irrigation in semi-aride tropical India*, Mimeo, ICRISAT, Patancheru. 1980.
[34] Fan, S. and Gulati, A., *Investment, Subsidies, and Pro-Poor Growth in Rural India*. IFPRI Discussion Paper 00716, International Food Policy Research Institute, Washington D.C., 2007.
[35] V. Ratna Reddy: *Land Degradation in India: Extent, Costs and Determinants.* Economic and Political Weekly, 38(44), pp. 4700-4713, 2003
[36] Goldsmith, E., Learning To Live With Nature the Lessons of Traditional Irrigation. *The Ecologist*, 28(3), 1. 1998.

Path dependence and critical junctures: an urban-rural water sharing study

L. Nicol
Department of Economics, University of Lethbridge, Canada

Abstract

Path dependence and critical junctures theory has evolved from first aiming to capture specific economic phenomena to explaining a host of social phenomena across a wide variety of disciplines. At this early stage in a planned research program, this paper will provide a literature review of path dependence and critical junctures. The subsequent study will focus on the history of urban-rural tension in the Calgary, Alberta region of Canada and the recent partnership that seeks to chart a new trajectory, which includes a vital water-sharing component. Paying particular attention to the potential access to water, the study will seek to determine whether powerful actors can vanquish an acrimonious path dependent past and carve out a new trajectory through what may potentially be a critical juncture.

Keywords: water management, path dependence, critical junctures, regional governance.

1 Background

There is a certain degree of urgency in improving water management in Alberta. A decade of unrelenting population and economic growth and the subsequent stress on water resources resulted in the termination of issuing new water allocations in 2005 in most of the southern part of the province. This means that existing water allocations need to be reallocated to new and existing users. Trading of water rights and water allocations is permissible under the province's *Water Act* and provides necessary flexibility. As irrigation controls 75% of all allocated water, it is expected irrigators will be an important part of the water reallocation process. Indeed a high-profile case occurred in 2007 involving the Western Irrigation District selling 2,500 dam^3 for a commercial development.

WIT Transactions on Ecology and the Environment, Vol 134, © 2010 WIT Press
www.witpress.com, ISSN 1743-3541 (on-line)
doi:10.2495/SI100161

However, the plebiscite held in the irrigation district narrowly passed and public opposition and an (unsuccessful) appeal to the Environmental Appeal Board underscored the high degree of discontent.

Against this backdrop, 17 municipalities in the Calgary region embarked on a bold initiative to develop a long-range coordinated approach to land use planning and water-sharing. The group ventured into new, untested and highly conflict-ridden territory. Resentment among the participants goes back decades as power over development shifted from urban control (under the Regional Planning Commissions of 1963) to individual municipalities (under the *Municipal Government Act* of 1995). In recent years as the population in the region boomed, individual municipalities forged ahead with development, uncoordinated with Calgary, thereby "escalating regional tensions" [1, p.618].

Despite this acrimonious history, a long-term regional planning process began in 2006 under the auspices of the Calgary Regional Partnership. After almost three years of work, the Calgary Regional Plan (CRP) was unveiled. The key feature of the CRP was the inclusion of density targets that would reduce urban sprawl and manage resources ecologically. The CRP represented a major breakthrough in water management in the province and a unique situation where water would be moving from the water-rich city of Calgary, to rural users. Unexpectedly, at a meeting of the general assembly of CRP members in June 2009, the three large and powerful rural municipalities voted against the CRP stating: "…it is clear that there is hangover amongst the CRP partners about old attitudes. The old rural versus urban clichés that have for so long created conflict instead of cooperation between neighboring municipalities continues to rear their ugly heads" [2].

Under the province's new Land Use Framework, regions are mandated to develop a regional plan. Thus, despite the disapproval by the three rural municipalities, the CRP was submitted to the Government of Alberta for a ruling. This ruling is expected in March, 2010, and will either impose the CRP on the region, or force the CRP members back to the table to resolve the outstanding issues.

Earlier legislation that created urban-rural tensions and current attempts to forge a new path represents a fruitful context for a study of path dependence and critical junctures. The Municipal District of Rocky View is the one large district of particular interest in this study as it has had the most acrimonious relationship with Calgary, yet is anxious to find new sources of water. Decades to ill-will appears to have become embedded and difficult if not impossible to reverse. However, the CRP presents an opportunity that may or may not succeed in setting the region along a new trajectory.

At this early stage in this research program, this paper will provide a literature review of path dependence and critical junctures. Given the research focuses on regional governance and municipal politics, attention will be given to their application to institutional change, broadly defined as structures and mechanisms of social order and cooperation governing the behavior of a set of individuals within a given human collectivity (Wikipedia). This work will ultimately help inform a more in-depth program of research. The paper is constructed as follows.

The proceeding section sets out the concepts of path dependence and critical junctures. The next section enumerates various criticisms of the concepts as well as potential solutions. Conclusions, including reflections on how path dependence and critical junctures can be used in the future study, are presented in the final section.

2 Path dependence and critical junctures

Although the concept of path dependence is a relatively recent phenomenon, only gaining prominence since the mid-1980s, its application has spread rapidly and broadly across many disciplines. What began as an innovative economic concept used to explain market failure has become ubiquitous, especially in social science research. Today some form of path dependence (select examples in brackets) is employed in economics [3–6]; politics and policy development [3, 7, 8], evolutionary economic geography [9–11], historical institutional development [12, 13] sociology [14–16] and management [17, 18]. With this widespread use came a myriad of interpretations and applications.

2.1 Path dependence

The genesis of path dependence resides in the discipline of economics with the seminal article by Paul David [19]. David showed how the emergence of touch-typing training programs during the 1890s created a demand for a standardized keyboard design. The "QWERTY" design held a marginal initial advantage over competing designs and became locked in, emerging as the industry standard because of this initial foot-hold despite later more efficient designs [20]. Generalized, this original concept of path dependence was associated with a type of market failure characterized by the early random adoption and subsequent lock-in of a relatively inefficient technology [19]. These processes are distinguished by technical interrelatedness, economies of scale and quasi-irreversibility of investments [21].

Path dependence is temporal, evolving over three distinct phases as described by Sydow et al. [22]. Phase 1 is characterized by a search process where choices are unconstrained. Once decisions are made, a dynamic self-reinforcing process is set into motion which eventually leads to a "deterministic pattern". This decision stage is labeled a critical juncture because once chosen, it becomes progressively difficult to return to the point where multiple options were available. Phase II is as a period where options are increasingly confined but still available. The final transition to Phase III is characterized by lock-in where viable options become even more constrained.

Lock-in is described as a property arising when "sequential patterns of activity form a 'groove' from which it subsequently becomes difficult to deviate" [23, p.36]. More specifically, a system can become 'over-committed' to a particular technology, industry or institutional regime, as the case may be [9]. Blyth notes that where policy continuity exists in a path dependent fashion, it may be a function of "cognitive locking" and as such, is an "ideational" rather than an institutional phenomenon [24, p. 23]. Wilsford [25] coined the phrase "behavioral lock-in" where certain behavior becomes embedded, gaining inertia and taking tremendous effort to change. Others characterize this state as being constrained by structures that confine and shape (institutions) and create lock-in through durable networks, which at the same time can result in instability when those networks change [25]. Weir spoke of "bounded innovation" where continuity is channeled but change is not entirely precluded [26]. Gulati [27] and Gulati and Gargiulo [28] observed that the path dependent character of inter-organizational networks means the previous ties among organizations increase the probability of an alliance between them in the future and that lock-in is likely to arise in the course of the development of the network [29].

Early on Arthur [30] augmented path dependence by applying it to more wide-spread technological adoption and the concept of increasing returns, coordination effects, adaptive expectations, and large setup costs that locked in particular technologies and excluded competitors [30]. Since then, Krugmen has focused on the way in which scale economies and positive externalities feed the process of industrial clustering while Arthur has gone on to study ways in which agglomeration externalities contribute to the concentration of firms in specific regions [31, 32]. Across a host of studies, regional development has been explained utilizing path dependence and related concepts (see for example [9]).

"Bygones are rarely bygones" [33, p.552]. In social dynamics, processes do not evolve in an unconditional way, but are "recursive" in the sense that former decisions have an impact upon those that follow [22], hence a path dependent process. In politics the notion of increasing returns was readily adopted by North [34] who argued that all the features identified by Arthur involving increasing returns in technology can be applied to politics. Institutions produce "massive" increasing returns in what North calls an "interdependent web of an institutional matrix" [34, p.95].

In a broad institutional context, new institutions often entail high start-up costs and considerable learning effects, coordination effects, and adaptive expectations [3]. Institutions "generate powerful inducements that reinforce their own stability and further development' [34. p. 255]. In politics, Pierson stressed, steps in a particular direction can trigger a self-reinforcing dynamic that accounts for sources of political stability. He further noted that once a particular path has been taken there are powerful incentives to stay on it [3]. Sydow et al. [22] stated that theories that point to practices or operational procedures in organizations, including concepts of "muddling through" [35], are related to the idea of path dependency. They stated: "These behavioral theories are sensitive towards the fact that history, as imprinted in existing routines and procedures for example, matter a lot in organizational behavior….Once underway, these routines and schemes shape decision-making processes in organizations and, eventually, cause organizational inertia…." [22, p. 12] Bednar and Page [36] argued that the best institutions for a particular society and its culture depend on previous institutional choices. The order in which institutions are implemented is significant, and existing institutions influence the performance of new institutions [37].

For Mahoney [37] and others like Pierson [3] the combination of some contingency at the front end and determinism at the back end of path dependent processes is what gives rise to what Thelen considers the most interesting features of path dependent social and political processes. These include unpredictability; sensitive dependence on initial conditions; non-ergodicity, where contingent events do not necessarily cancel out; and inertia, whereby once in motion, processes tend to stay in motion until an equilibrium or final outcome is reached (in [37]).

2.2 Critical junctures

Critical juncture has found wide-spread applicability in institutional research. Decisions made during such a juncture are deemed "critical" because such a decision leads to the creation of institutional or structural patterns that endure over time [37]. In studying nations' historic trajectories, Golob [38] applies the notion of "exogenous shock" such as a war or international economic crisis that serves as a critical juncture that sets a country onto a new path. Paths can, however, be "dislocated" by strong political leadership or in response to pressures such as climate change or rapidly rising fuel prices [12]. Krasner's [39] model, meanwhile, posits that institutional stability can be periodically "punctuated" by abrupt change, augmenting the notion of "punctuated equilibrium" developed by Kingdon [40] and Baumgartner and Jones [41]. So wide-spread is its applicability in research that the concept of critical junctures has been used to understanding the trajectories of national development in Latin America [42], to examining watersheds in American trade policy [43], to understanding the emergence of occupational welfare in Israel [44] (for extensive list see [45]).

In addition to the sources of change, scholars have contemplated the processes of change and the role of agency. Hogan and Doyle [4] believe that critical junctures which trigger significant change depend on "agents of change" – those who read a broad consensus upon, and subsequently consolidate around, one particular set of new ideas [5]. Garud and Karnoe [46] and Schumpeter [47] were first to challenge the notion that paths do not simply occur but can be deliberately brought about. Building on this argument, Sydow et al. [22] argued that agents can generate momentum for change over time. Rather than path dependence there is path creation and rather than increasing returns there is "positive self-reinforcing feedback" as well as social interaction of "more or less powerful agents" (p.28). In addition to these important agents, path breaking and path creating strategies are likely to require significant amounts of resources and "prolonged energy investment" (p. 34). Similarly, Torfing [48] spoke about providing more space for agency and social forces that carve out new trajectories. These he labels as "path shaping trajectories in a path-dependent context" that prevent old paths from reverberating [48, p.389].

Others have argued that institutional change occurs through gradual processes rather than abrupt critical junctures. Boas [13] claimed that in political institutions incremental changes can accumulate over time resulting in fundamental transformations. Thelen did not reject the concept of critical junctures and path dependence, but suggested most forms of institutional change involve endogenous, rather than exogenous, mechanisms of change through "layering" and "conversion" (in [49]). Hacker [50] extended this line of reasoning, coining "policy drift" as slow alterations due to changing socioeconomic circumstances.

3 Criticisms and potential solutions

Breznitz [51] defined and differentiated the myriad of path dependence approaches, all of which are, to him, confusing. As a result of constant broadening and overextending of the meaning of path dependence, the term has a "floating meaning" and runs the risk of becoming an "empty heading" [51, p. 28]. Path dependence has become, within a short period of time, a widely used concept in social science but, Greener noted, studies also have remarkably little in common in terms of their conceptual framework or approach [52]. The process of path dependence to date largely constitutes a research field with a just emerging and still incomplete theoretical framework [46].

Pierson argued that the greater range of path dependence study offered by loosening the definition has come at a high price in analytical clarity. It is a victim of what Sartori has called "conceptual stretching" [53, p. 1034]. Bridges [54] believed that path dependence loses its vigor outside its home base of economics because the central mechanism in economics, the market, does not function elsewhere. However, even in the realm of economics, path dependence has been problematic. As Martin and Sunley noted, almost from the instant that David and Arthur began writing about path dependence, interpretive debates erupted and are still unresolved [9].

Other issues include: distinguishing between difference degrees or types of path dependence, the relative difference between near and distant events and the relative significant of big versus small historical outcomes [9]. How to identify the critical juncture, or the starting point of the analysis, can be problematic. Mahoney [37] stated" "…the investigator is prone to keep reaching back in search of the foundational causes that underline subsequent events in the sequence" (p. 527).Others believe that path dependency is good at capturing constraint and inertia but not good at anticipating and explaining change [55]. Hess et al. [56] stated no empirical methods have been widely used that would allow general predictive capability, known as "path creation" in politics.

Equally important is the lack of criteria to assess a critical juncture compared to change that takes place incrementally [45]. As Mahoney stated, where does one draw the line as to what is and what is not a critical juncture? [37]. According to Sydow et al. [22], if we cannot define a critical juncture, the issue of path dependency becomes nebulous as we cannot ascertain the paths' origins, leaving the approach in a quagmire [22]. But developing rigorous standards for studying critical junctures would be exceedingly difficult given the vast array of topics that utilize the concept [22].

According to some, critical junctures have remained under examined and insufficiently specified [45]. As a result, the perspective from which critical junctures has been analyzed has varied greatly [45]. For example, the duration of a critical juncture may involve a relatively short period of time, for others, an extended period; it can focus on the underlying cleavages or crisis that precipitated a critical juncture or it can concentrate on the critical juncture itself [45].

The literature presents more issues than solutions to identifying and studying path dependence and critical junctures. A common definition is needed. Goldstone [57] suggested that path dependence be defined as:

> "….a property of a system such that the outcome over a period of time is *not determined* by any particular set of initial conditions….rather….one in which outcomes are related stochastically to the initial conditions and the particular outcome…depends on the choices or outcomes of *intermediate events* between the initial conditions and the outcome" (p. 834).

Boas [13] proposed a "composite-standard-model of path dependence" that incorporate the three components of increasing returns, conversion and layering that work as "cogs in a machine working together in a complex fashion to constitute real-world processes of continuity and change" (p.48). Graefe [58] said there is a need to consider how institutions are not just constrained but how actors or "path makers" intervene to make new trajectories possible. Hess et al. [56] linked the concept of discourse analysis or dominant speakers to path dependence and how their story lines that have shaped new policy. Indeed, discourses especially help explain reinforcing processes in politics because "…political discourses are always built on historically prior texts so that the past strongly determines future political actions" [56 p. 4].

Hogan [45] proposed a framework for determining a critical juncture. First, there must be a generative cleavage – a tension that leads to the critical juncture. This cleavage can stem from large-scale public dissatisfaction of less dramatic events such as urban-rural divides. Second, the change must be significant, swift and encompassing or enduring, i.e. having a legacy [45]. The issue of what constitutes "significant" will vary with the subject but when they occur, they have large and enduring consequences. Change is not, however, long and drawn-out and the duration of "swiftness" will depend on the topic and probably be arbitrary [45]. Finally, there must be a legacy; without it there is no critical juncture [45]. Hogan and Doyle [4] add a predictive element to critical junctures. For there to be a critical juncture, they must be a paradigm shift. This occurs where agents of change are able to reach a consensus on a new set of ideas [59], described by some as a "discursive phase"; the "institutional battle", where new ideas are "injected" into the policy domain, reducing barriers by coalition-building in this critical juncture [24, 57, 60].

4 Conclusions

Path dependence and critical junctures will have significant utility in gaining an understanding of the creation and persistence of animosity in the Calgary region that may prevent the creation of a critical juncture that could vastly improve water management in the region. The foundation of the study could focus on what appear to be two critical junctures - the passage of the *Municipal Government Act* and the CRP. Applying Hogan's [45] conceptual framework, discussed earlier, one must first, determine if there was a generative cleavage or a tension that led to the critical junctures. Hogan [45] in fact identifies urban-rural divides as possibilities. Second, the critical junctures would need to be proven to be significant, swift and encompassing or enduring. In the case of the CRP there must be a paradigm shift. Path dependence would have occurred between these first and second critical juncture and as such particular attention would need to be paid to intermediate events and how they are related to the initial conditions in the creation of urban-rural tensions, the institutions and the path makers that ultimately perpetuated that environment. Such institutions and path makers would be significant in charting a potential new path. In terms of methodology, discourse analysis of dominant players and their power to shape new policy may be particularly fruitful. Analysis of provincial archives will assist in understanding the issues that led to the passage of the 1995 legislation and the urban-rural tensions that preceded and followed it. Existing planning documents trace the CRP's decision-making process to date. This research will also analyze the dynamics of the ongoing CRP process as it continues to unfold.

References

[1] Ghitter, G. & Smart, A., (2009). Mad cows, regional governance, and urban sprawl. *Urban Affairs Review*, 44(5), 617-644.

[2] Calgary Regional Plan (CRP). General Assembly June, 19, 2009.

[3] Pierson, P., (2000). Increasing returns, path dependence, and the study of politics. *The American Political Science Review,* 94(2), 251-267.

[4] Hogan, J. & Doyle, D., (2007). The importance of ideas: An a priori critical juncture framework. *Canadian Journal of Political Science*, 40(4), 883–891.

[5] Altman, M., (2000). The behavioral model of path dependency: the economics of profitable inefficiency and market failure. *The Journal of Socio-Economics,* 29, 127-145.

[6] Barnes, W., & Stack, M., (2004). Old habits die hard: Path dependency and behavioral lock-in. *Journal of Economic Issues*, 38(2), 371-377.

[7] Hassink, R., (2005) How to unlock regional economies from path dependency? From learning region to learning cluster. *European Planning Studies,* 13(4), 521-535.

[8] Kay, A., (2005) A critique of the use of path dependency in policy studies. *Public Administration*, 83(3), 553–571.

[9] Martin, M., & Sunley, P., (2006). Path dependence and regional economic evolution. *Journal of Economic Geography*, 6, 395–437.

[10] Magnusson. L., (2001). The role of path dependence in the history of regulation. In L. Magnusson and J. Ottoson (eds) *The State, Regulation and the Economy: An Historical Perspective* (pp. 107-119). Cheltenham: Edward Elgar.

[11] Boschma, R & Lambooy, J.G., (1999). Evolutionary economics and economic geography. *Journal of Evolutionary Economics*, 9, 411-429.
[12] Low, N., & Astel, R., (2009). Path dependence in urban transport: An institutional analysis of urban passenger transportation Melbourne, Australia, 1956–2006. *Transport Policy,* (16), 47–58.
[13] Boas, T., (2007). Conceptualizing continuity and change the composite-standard mode. *Journal of Theoretical Politics,* 19(1), 33–54.
[14] Aminzade, R., (1992). Historical sociology and time. *Sociological Methods and Research*, 20, 462-467.
[15] Griffin L., (1992). Temporality, events and explanation in historical sociology: An introduction. *Sociological Methods and Research*, 20, 413-414.
[16] Somers, M. (1998). We're no angels: Realism, rational choice and relationality in social science. *American Journal of Sociology*, 104(3), 722-784.
[17] Kruger, A., 1996. The political economy of controls: America sugar. In: L. Alston and T. Eggertsson (eds) *Empirical Studies in Institutional Change* (pp.125-153). Cambridge: Cambridge University Press.
[18] Langlois, R.N., & Savage, D.A., (2001). Standards, modularity and innovation. In: R. Garud and P. Karnoe (eds*) Path Dependence and Creation (*pp.149-168). London: Lawrence Erlbaum Associated.
[19] David, P., (1985). Clio and the economics of QWERTY. *The American Economic Review*, 75(2), 332-337.
[20] Rast, J., (2009). Critical junctures, long-term processes, urban redevelopment in Chicago and Milwaukee, 1945–1980. *Social Science History* 33(4), 393-426.
[21] Martin, R., & Simmie, J., (2008). Path dependence and local innovation systems in city-regions. *Innovation: Management, Policy and Practice,* 10, 183-196.
[22] Sydow, J., Schreyogg, G., & Koch, J., (2005). Organizational Paths: Path Dependency and Beyond, second draft.
[23] Setterfield, M. (1997). Rapid Growth and Relative Decline: Modelling Macroeconomic Dynamics with Hysteresis. London: Macmillan.
[24] Blyth, M., (2001). The transformation of the Swedish model. Economic ideas, distributional conflict, and institutional change. *World Politics*, 54(1), 1-26.
[25] Wilsford, D. (1994) Path dependency, or why history makes it difficult but not impossible to reform health care systems in a big way. *Journal of Public Policy*, 14(3), 251-283.
[26] Weir, M., (1992). Ideas and the politics of bounded innovation. In S. Steinmo, K. Thelan, and F. Longstreth (eds) *Structuring Politics: Historical Institutionalism in Comparative Analysis* (pp.188-216). New York: Cambridge University Press.
[27] Gulati, R., (1995). Social structure and alliance formation. A longitudinal analysis. *Administrative Science Quarterly*, 40, 619-652.

[28] Gulati, R., & Gargiulo, M., (1999). Where do interorganizational networks come from. *American Journal of Sociology*, 104(5), 1439-1493.

[29] Gulati, R., Nohria, N., & Zaheer, A., (2000). Strategic networks. *Strategic Management Journal*, 21, 203-215.

[30] Arthur, W.B., (1989). Competing technologies, increasing returns, and lock-in by historical events. *Economic Journal*, 99, 116-131.

[31] Arthur, W.B., (1994). *Increasing Returns and Path Dependence in the Economy*. Ann Arbor: The University of Michigan Press.

[32] Krugman, P., (1991). *Geography and Trade*. Cambridge, Mass.: MIT Press.

[33] Treece, D.J., Pisano, G., & Shuen, A., (1997). Dynamic capabilities and strategic management. *Strategic Management Journal*, 18, 509-533.

[34] North, D.C., (1990). *Institutions, Institutional Change and Economic Performance*. Cambridge: Cambridge University Press.

[35] Lindblom, C.E., (1965). *The Intelligence of Democracy*. New York: Free Press.

[36] Bednar, J., & Page, S.E., (2005). Culture, institutional performance, and path dependence. Unpublished manuscript.

[37] Mahoney, J., (2000). Path dependence in historical sociology. *Theory and Society*, 29, 507-548.

[38] Golob, S., (2003) Beyond the policy frontier - Canada, Mexico, and the ideological origins of NAFTA *World Politics* 55, 361-98.

[39] Krasner, S.D., (1984). Approaches to the state: Alternative conceptions and historical dynamics. *Comparative Politics*, 16(2), 223-246.

[40] Kingdon, J., (1985). *Agendas, Alternatives and the Public Policies*. Boston: Little, Brown.

[41] Baumgartner, R.R., & B.D. Jones. (1993). *Agendas and Instability in American Politics*. Chicago: University of Chicago Press.

[42] Collier, R.B. & Collier, D., (1991). *Shaping the Political Arena: Critical Junctures, the Labour Movement, and Regime Dynamics in Latin America*. Princeton, N.J.: Princeton University Press.

[43] Haggard, S., (1988). The institutional foundations of hegemony: Explaining the reciprocal trade agreements Act of 1934. *International Organization,* 42(1), 91-119.

[44] Gal, J. & Bargal, D., (2002). Critical junctures, labour movements and the development of occupational welfare in Israel. *Social Problems*, 49(3), 432-454.

[45] Hogan, J., (2006). Remoulding the critical junctures approach. *Canadian Journal of Political Science,* 39(3), 657–679.

[46] Garud, R. & Karnoe, P., (2001). Path creation as a process of mindful deviation. In R. Garud and P. Karnoe (eds.) *Path Dependence and Creation* (pp. 1-38). New Jersey: Earlbaum, Mahwah.

[47] Schumpeter, J.A., (1942). *Capitalism, socialism and democracy*. New York: Harper & Row.

[48] Torfing, J., (1999). Towards a Schumpeterian workfare post national regime: path-shaping and path-dependency in Danish welfare state reform. *Economy and Society*, 28 (3), 369-401.

[49] Beland, D., (2007). Ideas and institutional change in social security: conversion, layering, and policy drift. *Social Science Quarterly*, 88, 20-38.

[50] Hacker, J., (2004). Privatizing risk without privatizing the welfare state: the hidden politics of social policy retrenchment in the United States. *American Political Science Review*, 98(2), 243-260.

[51] Breznitz, D., (2010). Slippery paths of (mis)understanding? Historically based explanations in social science. In G. Schreyogg and J. Sydow (Eds) *The Hidden Dynamics of Path Dependence Institutions and Organizations* (pp. 13-32). New York: Pagrave Macmillan.

[52] Greener, I., (2005). The potential of path dependence in political studies. *Politics* 25(1), 62–72.

[53] Sartori, G., (1970). Concept misformation in comparative politics. *The American Political Science Review*, 64(4), 1033-1053.

[54] Bridges, A. (2000). Path dependence, sequence, history, theory. *Studies in American Political Development*, 14, 109–112.

[55] Crouch, C., (2001). Welfare state regimes and industrial relations systems: the questionable role of path dependency theory. In B. Ebbinghaus and M. Manow (eds) *Comparing Welfare Capitalism: Social Policy and Political Economy in Europe, Japan and the USA* (pp. 105-124). London: Routledge.

[56] Hess, S., Kleinschmit, D., Theuvsen, L., Cramon-Taubadel, L., & Zschache, U., (2008). Conceptualizing path dependence through discourse analysis: The case of seasonal farm workers in Germany, *12th Congress of the European Association of Agricultural Economists*

[57] Goldstone, J.S., (1998). Initial conditions, general laws, path dependence and explanation in historical sociology. *American Journal of Sociology*, 104(3), 829-845.

[58] Graefe, P., (2004). Welfare regimes and the third sector: Rendering path dependency contingent? Paper presented to the 6th International Conference of the International Society for Third-Sector Research, Toronto, July.

[59] Hall, P.A., (1993). Policy paradigms, social learning and the state: The case of economic policy making in Britain. *Comparative Politics*, 25(3), 275-296.

[60] Oliver, M.J., & Pemberton, H., (2004). Learning and change in the 20th-Century British economic policy. *Governance*, 17(3), 415-414.

Modelling the adoption of different types of irrigation water technology in Alberta, Canada

S. Wheeler[1], H. Bjornlund[1,2], T. Olsen[1], K. K. Klein[2] & L. Nicol[2]
[1]*Centre for Regulation and Market Analysis, School of Commerce, University of South Australia, Australia*
[2]*Department of Economics, University of Lethbridge, Canada*

Abstract

This paper analyses farmers' adoption of hard and soft technology in relation to irrigation technologies, production changes and water management changes in Alberta, Canada. Greater significance was found in modelling the adoption of hard technology (such as irrigation infrastructure technologies) than modelling the adoption of soft technology (water management or irrigation area changes). Overall, some of the most important influences include farm size, irrigation technology, off-farm income and being a member of an irrigation district. Few socio-economic variables were found to be important. Adoption of soft technology most likely leads to greater water efficiencies and in the future greater attention should be paid to a wider variety of factors and influences in order to model water management and trading behaviour.
Keywords: hard and soft irrigation water technology, water management skills, water trading, Alberta, Canada.

1 Introduction

Due to increased water scarcity and escalating environmental problems across the globe there is an increased push to reduce water use and improve water use efficiency, especially for irrigation [1]. Some areas in Canada have faced increasing water shortages since the early 1990s, with irrigators traditionally being the largest water consumer. The majority of irrigated land in Canada is in Alberta (64%), with most of this irrigation in the South Saskatchewan River Basin (SSRB) and it accounts for 71% of consumptive use of surface water [2].

WIT Transactions on Ecology and the Environment, Vol 134, © 2010 WIT Press
www.witpress.com, ISSN 1743-3541 (on-line)
doi:10.2495/SI100171

The majority of irrigators in southern Alberta are members of organized irrigation districts. Those who are members in irrigation districts account for 82% of total irrigators and about three-quarters of land under irrigation. They pay a flat fee per hectare to the irrigation districts to cover their share of administration, maintenance and some infrastructure rehabilitation costs. Irrigation districts hold the water licenses and manage the supply infrastructure. Private irrigators have individual water licenses and obtain water from various rivers within the SSRB; they are responsible for all infrastructure needed to pump the water from the river and convey it to their fields as well as the irrigation infrastructure in the fields [3]. Neither type of irrigator pays for the water, the cost of head works and the supply infrastructure. Water licenses are tied to the land and, historically, have remained with the property when the land was sold [4].

Although water use efficiency in Alberta has increased over time thanks to the adoption of more efficient water use technologies (a movement away from surface irrigation to wheel-move and then to high pressure and low pressure centre pivots), water resources in the SSRB have been found to be fully or over-committed, with pressure on the water resource expected to worsen over time due to predicted population and economic growth as well as increased environmental demand and climate change. Improving irrigation equipment and water management practices can reduce the need for water.

It is important to know which factors influence the adoption of improved irrigation technologies on farms before an effective policy can be designed. The purpose of this study is to determine the influences on the irrigation technology and management practice adoption in Alberta, in particular focusing on the difference between the adoption of hard and soft water technologies.

1.1 Adoption of new technologies in agriculture

Many studies have investigated the factors that influence farmers' adoption of new technologies [5-9]. The general findings have been that early adopters of new technologies tended to be younger, more educated, more cosmopolitan, have higher incomes, larger farm operations, are leaders within the community, and more reliant on primary sources of information [10]. The expected profitability of new technologies also has consistently been shown to be linked with adoption (as first shown by Griliches [5]).

New technologies can be of two distinct types: hard and soft technologies (also described as embodied and disembodied technologies). Hard technologies are defined as those innovations where adoption involves buying/installing a physical input or output, while soft technologies involve the farmer acquiring new skills or rearranging their farm management or decision making in some way. These technologies can be easier than others for farmers to adopt (information is more readily available) because they involve a physical tangibility that can be experienced immediately. On the other hand, learning new skills or changing farmers' mindsets about management issues may be influenced by different sociological factors [11] and hence adoption is more difficult to predict. Data on soft technologies are often more difficult for farmers to obtain because there are fewer commercial benefits involved and, hence, reduced marketing and information provided by agricultural organisations. This is especially true for innovations with high environmental benefits and may also be true for changing water management practices that involve a range of private and public benefits. Water management practices can involve the implementation of catchment-wide practices that have a range of externalities associated with them.

In this study, we distinguish between different forms of hard and soft water technologies. The forms of hard water technology discussed here include surface irrigation, wheel move irrigation, pivot irrigation (with high and low pressure systems) and the use of computer panels to operate the pivots. The adoption of more efficient irrigation systems allows a higher proportion of water to be applied to crop root zones, hence increasing the level of water consumption for crops at a given level of water application. As such, irrigators can meet the consumptive needs of the plants while reducing their overall water use, thereby increasing their water use efficiency [12]. It is important to note that our definition of hard water technology traditionally falls more into a soft technology grouping in the agricultural adoption literature. This is because irrigation scheduling systems and water efficiency technologies increasingly are promoted with an environmental rationale, with the idea that they would reduce water usage within the catchment area. Because of these benefits, they would often be described as soft technologies in the literature. However, the water saving argument from adoption of new irrigation technologies is far from certain, as discussed later. The forms of soft water technology include production changes (increasing/decreasing irrigation); soil monitoring assessment (using visual crop condition, hand auger and feel method, or soil moisture instruments); irrigation scheduling techniques (use of irrigation or climate models); and the use of private consultants.

There has been considerable research on the adoption of irrigation technologies on farms [1, 12-21]. Scheierling et al. [22] found irrigation technologies were more likely to be adopted when adoption costs were lower, on land of lower quality and on crops of higher quality. Moreno and Sunding [15] found that drip technology was adopted on land with a higher gradient and poorer water retention capacity. On the other hand, Zhou et al. [13] found adoption of water saving technology higher on land with better water holding capacity.

Increasing the price of water has been identified as a key incentive for irrigators to adopt water-saving irrigation systems [15]. Their findings were supported by Carey's and Zilberman's [23] model which showed that random events like drought encouraged farmer adoption of modern irrigation technologies. Lower quality water also has been found to lead to adoption of more technically efficient irrigation methods [12]. However, the existing literature on the adoption of improved irrigation technologies tends to emphasize price as an indicator of scarcity rather than actual measures of physical scarcity.

Much of the relevant literature on irrigation technologies has been based on conditions in Israel or the United States, and there has been less research in

Canada. On the other hand, there have been comparatively fewer studies on changes to water management in the literature [e.g. 24-25]. This limited literature suggests that the adoption of best management irrigation is not widespread and there is considerable capacity for improvements in practices. Deng et al. [14] investigated the challenges for water saving in agriculture in China, and highlighted the need to breed new varieties for high water use efficiency. As the review by Pereira et al. [26] on water irrigation management under water scarcity indicated, there is a need to develop appropriate methodologies for the analysis of social, economic, and environmental benefits of improved irrigation management.

2 Methods

2.1 Data and methodology

The data used in this study were obtained from two surveys, one of private and one of district irrigators in southern Alberta. Telephone interviews were conducted with 150 private irrigators in March 2007 [5]. Irrigators were randomly selected from a list of names and locations provided by Alberta Environment. The survey of farmers in the irrigation districts was conducted in 2006, with questionnaires sent by mail to 810 irrigators (320 in the Raymond Irrigation District and 490 to the Taber Irrigation District) [5]. 150 questionnaires were returned (21% and 17% response rate from Raymond and Taber, respectively).

Both surveys asked for information on past, current and planned initiatives in adoption of new irrigation technologies, production, water management and energy use (and barriers to further improvements), in three distinct time periods: historical (prior to 2001), recent past (2001-2006) and future (2007-2012).

Binary probit analysis was used to examine the factors influencing adoption of past and future irrigation technologies and management practices, classified into hard or soft technologies. Binary probit models are appropriate when the dependent variable to be evaluated is dichotomous. If a farmer had adopted a form of technology/management change prior to 2006 (i.e. they had either adopted prior to 2001 or from 2001 to 2006 in the survey categories), then these farmers were categorized as adopters and were coded as 1, and other farmers were categorized as non-adopters and coded as 0.

Five broad categories of irrigator adoption of hard technologies were modelled, namely: a) conversion from surface to wheel move; b) conversion from wheel move to pivot; c) conversion from surface to pivot; d) conversion from high pressure to low pressure pivots; and e) purchase of a computer panel for pivot. Water management (soft technology) questions included: a) monitor soil moisture using visual crop condition or hand auger and feel method or start using soil monitoring instruments; b) start to turn water on and off and/or monitor position of pivots using computer or phone; c) start to use Alberta Irrigation Management Model (AIMM) or Irrigation Management Climate Information Network (IMCIN) to schedule irrigation; and d) start to use private consultants to support irrigation decision making.

Table 1: Variable summary.

Abbreviation	Explanation
Surfacewheel	Convert from surface to wheel move irrigation, 1 = adoption, 0 non-adoption
Wheelpivot	Convert from wheel move to pivot irrigation, 1 = adoption, 0 = non-adoption
Surfacepivot	Convert from surface to pivot irrigation, 1 = adoption, 0 = non-adoption
Highlow	Convert from high to low pressure irrigation system, 1 = adoption, 0 = non-adoption
Comppiv	Purchase computer panel for pivot, 1 = adoption, 0 = non-adoption
Visualcrop	Monitor soil moisture using visual crop condition, where 1 = adoption, 0 = non-adoption
Handauger	Monitor soil moisture using hand auger and feel, 1 = adoption, 0 = non-adoption
Soilmonitor	Monitor soil moisture using soil monitoring instruments, 1 = adoption, 0 = non-adoption
Compphpiv	Use computer or phone to monitor pivot position, 1 = adoption, 0 = non-adoption
AIMM	Use AIMM or IMCIN to schedule irrigation, 1 = adoption, 0 = non-adoption
Consult	Use private consultants to support irrigation decision making, where 1 = adoption, 0 = non-adoption
Own	1=owner of irrigated land; 0 otherwise
Age	25=under 34 years; 45=between 35 and 54 years; 65=over 55 years
Farmy	5=has been involved with farming less than 10 years; 17=involved between 11 and 24 years; 30=involved more than 25 years
Parent	1=parents in farming; 0 otherwise
Child	1=expect a member of family to take over farm; 0 otherwise
edu	9= not completed high school; 12=completed high school; 15= undergraduate university degree or college degree; 17= postgraduate degree
Offfarm	1=a member of the household had off-farm work; 0 otherwise
offarminc	The percentage of income derived from off-farm work
dryland	Number of dryland acres on farm
irrig	Number of irrigated acres on farm
Priv	1=private irrigator; 0=district irrigator
Qual	1=farmer is concerned about irrigation water quality; 0 otherwise
Pctcer	Percentage of irrigated acres producing cereals
Pctsc	Percentage of irrigated acres producing speciality crops
Pctgrav	Percentage of acres irrigated by gravity
Pctwm	Percentage of acres irrigated by wheel move
Pctlpp	Percentage of acres irrigated by low pressure pivot
Pcthpp	Percentage of acres irrigated by high pressure pivot
Pctele	Percentage of acres irrigated using electricity
Pctng	Percentage of acres irrigated using natural gas
Pctdie	Percentage of acres irrigated using diesel
Pctpro	Percentage of acres irrigated using propane

Estimation of the binary probit model was based on maximum likelihood. The estimated coefficients represent the impact of a one-unit change in the independent variable in question, holding the other explanatory variables constant, on the log of the probability of a given choice, not on the probability itself. Definitions of all variables are provided in Table 1.

A range of farmer socioeconomic demographic variables, farm characteristics, current irrigation practices (irrigation types and energy uses) and farmer beliefs/attitudes were included in our models. In the models explaining the adoption of new irrigation technologies, ownership of irrigated land is expected to have a positive impact on adoption since the returns to investment mainly accrue to land owners.

As physical hard technologies tend to be long-run investments; the sign of the estimated owner coefficient is expected to be positive; while soft water management practices are skills based, transferable and hence are more like short-term investment, the owner coefficient is expected to be negative.

In general, the older a farmer is, and the longer the farmer has been farming, the slower the farmer is expected to be in adopting new innovations. However, the latest anecdotal evidence from Alberta suggests that older farmers adopt newer irrigation technology because older types of irrigation such as flood and wheel move require too much physical activity, hence it is expected that there will be an opposite difference between the age coefficient effect in the hard and soft technology models. It is also expected that if the farmer took over the farm from a parent, the farmer holds a traditional view on farming, which is a negative influence on the adoption of new technologies and management practices. The opposing argument is that if the farmer expects his/her children to take over the farm, adoption of new technologies should be greater because the benefits of the investments are kept in the family. It is expected that the more education a farmer has, the more likely he/she is to adopt new technology.

Off-farm income may pull adoption in both directions. Off-farm income provides an opportunity to finance technology investment and also affords more time for off-farm work as the new technology generally is less labour intensive. But, high off-farm income makes the farm less reliant on farm income and therefore the farmer may be unable to invest in new farm technology. The larger the size of the farm, the higher the expected returns, therefore the size of farm is expected to have a positive impact on adoption, though the size of the irrigated area should have more of an impact than the farm's dryland area. If a farmer is concerned about the quality of irrigation water, it is expected that the farmer is more likely to invest in new technologies that improve irrigation water management. The concern about water quality is a proxy for environmental awareness inducing more efficient water use. The coefficient for private is expected to be negative as irrigation within district is more intensive.

3 Results

Results are shown in Tables 2 and 3. Although the probit regression models predict only moderately well, they exhibit no problems with collinearity (though

Table 2: Probit regressions of hard technology: irrigation equipment adoption (1 = adoption, 0 = non-adoption) (n = 269).

	Surface to wheel move		Wheel move to pivot		Surface to pivot irrig.		High to low pressure		Computer panel pivot	
	Coef.	P>|z|	Coef.	P>|z|	Coef.	P>|z|	Coef.	P>|z|	Coef.	P>|z|
own	-0.08	0.76	-0.22	0.39	-0.01	0.98	-0.15	0.57	-0.74**	0.02
age	0.00	0.78	0.02*	0.08	0.02	0.21	-0.01	0.31	-0.01	0.34
Farmy	0.02	0.26	-0.01	0.49	0.04	0.13	0.02	0.32	0.02	0.35
parent	-0.34	0.23	0.05	0.88	-0.48	0.18	0.82**	0.02	0.21	0.58
Child	-0.01	0.95	-0.21	0.26	-0.00	0.99	-0.01	0.96	-0.34*	0.10
Edu	-0.02	0.52	-0.03	0.42	0.00	0.97	-0.04	0.28	-0.15***	0.00
offfarm	0.26	0.34	0.76***	0.01	-0.28	0.51	-0.09	0.76	0.01	0.97
offarminc	-0.01**	0.05	-0.01***	0.00	-0.01	0.42	0.01	0.30	0.00	0.45
dryland	0.00	0.13	-0.00	0.17	0.00	0.13	-0.00	0.13	-0.00	0.77
Irrig	0.00	0.48	-0.00***	0.01	0.00	1.00	-0.00	0.580	0.00	0.37
Pctcer	-0.01	0.12	-0.00	0.41	-0.00	0.89	-0.01	0.24	-0.01	0.25
Pctsc	-0.00	0.46	-0.00	0.48	0.00	0.87	0.01*	0.07	0.01	0.17
pctgrav	-0.01	0.29	0.01	0.24	-0.01	0.26	-0.00	0.51	-0.00	0.64
Pctwm	0.00	0.63	0.01*	0.06	-0.00	0.48	-0.00	0.68	0.00	0.68
Pctlpp	-0.01	0.26	0.03***	0.00	-0.00	0.52	0.01	0.15	0.01	0.44
pcthpp	-0.01	0.34	0.02***	0.00	0.01	0.42	0.00	0.70	0.01	0.34
Pctele	-0.00	0.21	0.00	0.37	0.01*	0.06	0.01**	0.03	0.01**	0.02
Pctng	0.00	0.89	0.00	0.71	0.00	0.58	0.01	0.12	0.00	0.89
Pctdie	-0.00	0.31	-0.00	0.28	0.01	0.34	-0.01	0.3	-0.00	0.90
Pctpro	-0.00	0.72	-0.00	0.58	0.01	0.37	0.01	0.25	-	-
Qual	0.26	0.16	-0.20	0.28	0.51**	0.05	0.24	0.23	0.01	0.97
Priv	-1.06***	0.00	-1.06***	0.00	-0.84***	0.01	-0.97***	0.00	-0.54*	0.07
_cons	0.33	0.72	-1.42	0.20	-2.6	0.03	-0.75	0.44	0.43	0.69
Waldchi2	49.72***		87.43***		44.26***		61.68***		57.11***	
PseudoR2	0.147		0.25		0.21		0.22		0.21	
Log pseudolikelihood	-141.09		-127.05		-71.67		-117.7		-90.64	

Table 3: Binary probit regressions of soft technology: water management changes (adoption = 1, non-adoption = 0) (n = 269).

	Visual crop cond.		Hand auger/feel		Soil monitor		Computer/phone use		AIMM/IMCIN		Private consultants	
	Coef.	P>|z|	Coef.	P>|z|	Coef.	P>|z|	Coef.	P>|z|	Coef.	P>|z|	Coef.	P>|z|
own	-0.45*	0.06	-0.10	0.70	-0.55	0.22	-0.69	0.13	0.24	0.49	0.27	0.30
age	0.01	0.29	0.01	0.64	0.00	0.92	-0.00	0.96	-0.03***	0.01	0.00	0.67
farmy	0.00	0.96	0.00	0.89	-0.00	0.87	0.02	0.42	0.02	0.40	0.04**	0.03
parent	-0.16	0.55	0.62**	0.05	-0.07	0.84	-0.37	0.30	-0.27	0.54	0.14	0.63
child	0.01	0.96	-0.19	0.31	0.03	0.90	-0.35	0.18	0.30	0.21	-0.43**	0.03
edu	0.04	0.30	0.05	0.19	0.03	0.51	-0.06	0.24	0.01	0.92	0.03	0.51
offfarm	0.16	0.57	0.39	0.16	0.36	0.32	0.08	0.84	0.45	0.19	0.80***	0.01
offfarminc	-0.01	0.17	-0.00	0.68	0.00	0.42	0.01	0.40	-0.00	0.70	-0.01	0.12
dryland	-0.00**	0.03	-0.00**	0.04	0.00**	0.02	0.00**	0.05	-0.00***	0.01	0.00	0.97
irrig	-0.00	0.62	0.00	0.86	0.00	0.62	0.00	0.79	0.00*	0.09	0.00	0.11
pctcer	0.01***	0.00	-0.00	0.91	0.00	0.64	-0.00	0.34	0.00	0.92	-0.00	0.61
pctsc	-0.00	0.41	0.01*	0.09	0.01**	0.03	0.00	0.69	0.01	0.30	0.00	0.29
pctgrav	-0.00	0.98	0.00	0.52	0.00	0.51	-0.00	0.69	0.00	0.54	-0.01	0.33
pctwm	-0.00	0.94	0.01	0.11	0.01**	0.05	0.00	0.98	0.01**	0.05	0.00	0.85
pctlpp	-0.01*	0.06	0.02***	0.00	0.02***	0.00	0.00	0.85	0.00	0.46	0.00	1.00
pcthpp	-0.01	0.15	0.02***	0.00	0.02***	0.00	0.01	0.32	0.01	0.17	0.01*	0.08
pctele	0.00	0.43	-0.00	0.37	0.01**	0.05	0.00	0.40	0.01*	0.10	0.00	0.59
pctng	0.00	0.66	-0.01*	0.07	0.01	0.13	-0.01	0.19	0.00	0.72	0.00	0.85
pctdie	-0.00	0.41	-0.01***	0.00	0.01**	0.03	0.00	0.67	0.01	0.19	-0.00	0.91
pctpro	0.00	0.73	-0.01	0.11	0.01	0.27	-	-	-	-	0.01	0.38
qual	-0.03	0.88	0.40**	0.03	-0.03	0.89	-0.01	0.98	-0.06	0.83	0.13	0.50
priv	0.36	0.18	-0.14	0.59	0.22	0.46	-0.28	0.34	0.33	0.27	0.45	0.13
_cons	0.03	0.97	-2.59***	0.01	-4.35***	0.00	-1.42	0.28	-1.73*	0.09	-3.02***	0.00
Waldchi2	53.64***		88.7***		47.53***		55.42***		42.07***		42.99***	
PseudoR2	0.153		0.281		0.199		0.158		0.188		0.178	
Logpseudo-likelihood	-142.74		-131.88		-70.25		-48.73		-57.85		-108.62	

a number of variables were dropped initially from some models) and all were estimated with robust standard errors to minimize heteroscedasticity issues. There was more success achieved in modelling the adoption of hard technologies (i.e., actual equipment) rather than the adoption of soft technologies (changing water management practices). In general (where a variable was at least significant twice in Table 2 with the same sign), irrigators were more likely to have adopted new irrigation technologies if they: a) had a lower percentage of off-farm income as a source of total farm income; b) had a higher percentage of land irrigated by electricity; and c) were a member of an irrigation district. In general they were more likely to have adopted water management practices if they: a) had less dryland acres; b) had a higher percentage of land devoted to speciality crops; c) had a higher percentage of area irrigated by wheelmove or low (or high) pressure pivot using electricity.

In the tables, a *** signifies significant at least at the 0.01 level, a ** signifies significant at least at the 0.05 level and a * signifies significant at least at the 0.10 level. Similar low significance levels were achieved when modelling increases or decreases in the total area under irrigation (results not shown here). Irrigators were more likely to have increased their irrigated area if they had a lower percentage of area irrigated by gravity. Irrigators were more likely to have decreased their irrigated area if they: a) had a larger dryland area; b) had a smaller irrigated area; c) had a higher percentage of area with speciality crops; d) had a higher percentage of area irrigated by gravity, wheelmove, low or high pressure pivot; and e) were not concerned with the irrigation water quality. Irrigators were more likely to have adopted less water using crops (results not shown here) if they: a) did not have off-farm income (but the higher the percentage of off-farm income then the more likely that adoption would have occurred); b) had a larger irrigated area; c) had a higher percentage of area irrigated by wheelmove and high pressure pivot; d) had a lower percentage of area irrigated by electricity; and e) were not concerned with the irrigation water quality.

4 Discussion

These results confirm that modelling the adoption of skill-based or sustainable management activities requires careful attention to a variety of influences. The traditional variables used in economic models do not sufficiently explain farmers' actions and intentions. Greater consideration needs to be given to other factors, such as farmers' attitudes to their lifestyle, their farm desires, succession, environmental factors and community views. Perhaps there is an endowment or ownership utility effect attached to farm production, with greater disutility experienced by reducing production overall rather than increasing production.

One of the most consistently significant factors affecting the adoption of new water management technologies was whether the irrigator was private or a member of an irrigation district. Private irrigators clearly invest less in hard technologies, most likely because they rely less on high value irrigation crops to generate farm revenue, but rather use irrigation to produce feed to supply their main businesses such as feedlot or cow-calf operations. Generally, the larger the dryland area of the farm, the less likely it is that an irrigator adopted hard water technologies, although it has a positive effect on the adoption of some water management techniques. This suggests that the returns to the different water investments do depend to some extent on the size of the farm.

The next most important influences were the type of technologies already present on the farm. Farms that had already invested heavily in various irrigation technologies such as high or low pressure pivot were more likely to adopt improved water management practices. This is likely to reflect that these irrigators, already having the most efficient forms of hard water technology, are looking at soft technologies to further increase their efficiency. On the other hand, it is irrigators who have a higher percentage of their land irrigated by electricity who are more likely to adopt hard water technologies.

The next most important influence was if the farm had off-farm income, and the percentage of total household income derived from off-farm sources. The effect on technology adoption of earning off-farm income was not clear since both income and substitution effects affect it. The results show that having off-farm income increases the probability of adopting some hard and soft technologies (income effect) but having a larger percentage of total farm income derived from off-farm work can decrease adoption of some hard technologies (substitution effect). This indicates that when farmers have some off farm work they have an interest in being more efficient as it saves time; however, it seems that the higher the dependence on off-farm work the weaker their overall farm financial position and hence their ability to invest.

There were few socio-economic influences on the adoption of improved technologies. As hypothesized, younger farmers were more likely to have adopted soft water technology (such as sophisticated irrigation management models) as compared to older farmers who were more likely to adopt hard technologies for increased convenience (i.e. move away from wheel move to pivot). Education did not play that much of a role except in negatively influencing the adoption of one hard technology. Given that this result was unexpected, sensitivity testing of our education variable with dummy education variables found that having a graduate degree had a significant, positive effect on the adoption of soft technologies, reinforcing our conclusion of the importance of further education as a driver of adoption of more skilled-based technologies.

5 Conclusion

The need to improve irrigation water use efficiency and conserve water is a key issue within the South Saskatchewan River Basin in Alberta, Canada. Water use efficiency in the region may be improved by: a) improving irrigation equipment; b) improving water management practices; and c) reducing the irrigated area and/or changing production to less water consuming varieties or crops. Before any effective government policy on water use efficiency is designed, it is important to know what actually influences the adoption of improved irrigation technologies and how the adoption differs between different categories of hard and soft technologies. Hard technology was defined as those innovations where adoption involves buying/installing a physical input or output, while soft technology involves the farmer acquiring new skills or rearranging farm management in some way. The most important way that improved water use efficiencies is likely to occur is from adoption of soft technologies, in terms of reducing the area irrigated, improving irrigation scheduling, switching to less water consuming crops or varieties, or implementing water trading. Although no policy incentives may be required to increase the adoption of hard water technologies, given the primarily private benefits that accrue to irrigators, a range of instruments such as education, public extension and subsidy approaches may be necessary to encourage further adoption of soft water technologies.

Our results revealed that more significance was achieved in the models of hard technology (specifically, improved irrigation equipment) than soft technology (improving management practices and modifying irrigation/production areas), suggesting that the adoption of skill-based or sustainable management practices requires careful attention to a variety of influences. Greater consideration needs to be given to other factors, such as farmers' attitudes to their lifestyle, succession, environmental factors, community views, information sources, institutional influences, etc.

The most important influence on adoption seems to be the farms dependence on irrigation. District irrigators with a large proportion of their irrigated area in specialty crops and hence are very dependent on sufficient and timely irrigation are the ones most likely to adopt. On the other hand private irrigators with large areas of dry land farming and small irrigated areas are least likely to adopt as they are least dependent on irrigation. Having off-farm income and its importance in the overall farm total income had both an income and substitution impact on technology and skill adoption. Few socio-economic influences were found to significantly affect the adoption of hard or soft technologies.

References

[1] Seo, S., Segarra, E., Mitchell, P., & Leatham, D., Irrigation technology adoption and its implication for water conservation in the Texas High Plains: a real options approach. *Agricultural Economics,* 38(1), 47-55, 2008.

[2] AENV, Alberta Environment. 2003. W*ater for life – Alberta's strategy for sustainability,* Alberta Environment, Edmonton.

[3] Bjornlund, H., Nicol, L. & Klein, K., Economic instrument and irrigation water management – a comparative study of private and district irrigators in Alberta, Canada. *In Sustainable Irrigation – Management Technologies and Policies II*, ed. Y. Esteve, C. Brebbia and C. Rico, pp. 3-14,2008.

[4] Bjornlund, H., Nicol, L. & Klein, K., Challenges in implementing economics instruments to manage irrigation water on farms in Southern Alberta. *Agricultural Water Management*, 92(3), pp.131-41, 2007.

[5] Griliches, Z. Hybrid Corn: An exploration in the economics of technological change. *Econometrica*, 25(4), pp.501-21, 1957.

[6] Feder, G. & Slade, R., The acquisition of information and the adoption of new technology. *American Journal of Agricultural Economics*, 66, pp. 312-320, 1984.

[7] Feder, G. & Slade, R., The Role of Public Policy in the Diffusion of Improved Agricultural Technology. *American Journal of Agricultural Economics*, 67(2), pp.423-28, 1985.

[8] Guerin, L. & Guerin, T., Constraints to the adoption of innovations in agricultural research and environmental management: a review. *Australian Journal of Experimental Agriculture*, 34(4), pp. 549-71, 1994.

[9] Rogers, E., *Diffusion of Innovations*, 5th ed., The Free Press: New York, 2003.

[10] Stephenson, G., The somewhat flawed theoretical foundation of the extension service. *Journal of Extension*, 41(4), pp.1-10, 2003.

[11] Morrison, M., Identifying market segments for technology adoption. 49th Conf. of the Australian Agricultural and Resource Economics Society, Coffs Harbour, NSW, 9-11th Feb 2005.

[12] Schuck, E., Frasier, W., Webb, R., Ellingson, L. & Umberger, W., Adoption of more technically efficient irrigation systems as a drought response. *Water Resources Development*, 21(4), pp. 651-62, 2005.

[13] Zhou, S., Herzfeld, T., Glauben, T., Zhang, Y. & Hu, B., Factors affecting Chinese farmers' decisions to adopt a water-saving technology. *Canadian Journal of Agricultural Economics*, 56(1), pp. 51-61, 2008.

[14] Deng, X., Shanc, L., Zhang, H. & Turner, N., Improving agricultural water use efficiency in arid and semiarid areas of China. *Agricultural Water Management*, 80(1-3), pp.23-40, 2006.

[15] Moreno, G. & Sunding, D., Joint estimation of technology adoption and land allocation with implications for the design of conservation policy. *American Journal of Agricultural Economics*, 87(4), pp. 1009-19, 2005.

[16] Schuck, E. & Green, G., Field attributes, water pricing, and irrigation technology adoption. *Journal of Soil and Water Conservation*, 56(4), pp. 293–98, 2001.

[17] Green, G., Sunding, D., Zilberman, D. & Parker, D., Explaining irrigation technology choices: a microparameter approach. *American Journal of Agricultural Economics*, 78(4): pp.1064–72, 1996.

[18] Caswell, M. & Zilberman, D., The choices of irrigation technologies in California. *American Journal of Agricultural Economics*, 67(2), pp.224–34, 1985.

[19] Caswell, M., Irrigation Technology adoption decisions: Empirical evidence. In *The Economics and Management of Water and Drainage in Agriculture*, ed. A. Dinar & D. Zilberman, Kluwer Academic Publishers, Boston, pp.295-312, 1991.

[20] Shrestha, R., & Gopalakrishnan, C., Adoption and Diffusion of Drip Irrigation Technology: An Econometrics Analysis. *Economic Development and Cultural Change*, 41(2), pp.407– 18, 1993.

[21] Negri, D. & Brooks, D., Determinants of irrigation technology choice. *Western Journal of Agricultural Economics*, 15(2), pp.213–23, 1990.

[22] Scheierling, S., Loomis, J. & Young, R., Irrigation water demand: A meta-analysis of price elasticities. *Water Resources Research*, 42(1), W01411, doi:10.1029/2005WR004009, 2006.

[23] Carey, J. & Zilberman, D., A model of investment under uncertainty: modern irrigation technology and emerging markets in water. *American Journal of Agricultural Economics*, 84(1), pp. 171-83, 2002.
[24] McCrea, R. & Rivers, M., Sustainable irrigation – a collective effort for regional development. Proc. of the Int. Conf. of the Network of Regional Governments for Sustainable Development. Fremantle, Western Australia, 2003.
[25] Bauder, T., Waskom, R., Frasier, M. & Hoag, D., Irrigation best management practices: What are Colorado producers using? Fact sheet #19, Colorado State University, 1997.
[26] Pereira, L., Oweis, T. & Zairi, A., Irrigation management under water scarcity. *Agricultural Water Management*, 57(3), pp.175-206, 2002.

Improving water use efficiency in southern Alberta irrigated agriculture: choice of criterion

K. K. Klein[1], S. N. Kulshreshtha[2], M. K. Ali[1] & R. Bewer[1]
[1]*Department of Economics, University of Lethbridge, Canada*
[2]*Department of Bioresource PBE, University of Saskatchewan, Canada*

Abstract

Irrigation accounts for the largest consumption of water in many areas of the world, including Alberta. As water supplies dwindle under the threat of climate change, several environmental lobby groups have argued that less water should be available for irrigation with the hope that more water would then be available for environmental objectives. The agricultural industry may be able to counteract these criticisms if its water use efficiency could be increased and substantiated. This requires selection of an appropriate criterion to use for measuring improvement in water use efficiency (WUE). In this study, two aspects of WUE are examined: Technical and Economic. Estimation is based on data for two river sub-basins where irrigation is practised in Alberta, using data over a five-year period, 2004-2008. The Oldman river sub-basin covers a larger area, and the amount of water diverted reflects it. Water use technical efficiency for this sub-basin was also higher than that for the Bow river sub-basin. Similar results were obtained for water use economic efficiency. On average, in the Oldman river sub-basin, one dam^3 (1000 cubic meters) of irrigation water generates 4.4 tonnes of production, as against only 2.7 tonnes for the Bow river sub-basin. In terms of economic efficiency, water applied contributes on average, \$171 per dam^3 for the Oldman versus only \$115 for the Bow river sub-basin. Final assessment of these measures suggests that the physical efficiency measures suffer from the limitation of ignoring the human aspect of efficiency improvement. Thus, improving the technical efficiency may not necessarily lead to improvement in human well-being, which is one of the major objectives of

www.witpress.com, ISSN 1743-3541 (on-line)
doi:10.2495/SI100181

irrigation development. Economic measures are, therefore, expected to provide a better measure of change in the human welfare and are recommended for future use, although even these can be improved further.
Keywords: water use efficiency, irrigation, Alberta, technical efficiency, economic efficiency.

1 Background

1.1 Need for irrigation in Alberta

To a great extent climate factors play a major role in determining the type and extent of agricultural production in Canada. According to Shady [1], Southern Alberta is one of the regions in Canada where average water deficits are in excess of 200 mm, partly as a result of the dry and sunny climate in general. Dale-Burnett [2] described the region as the 'Palliser Triangle', (named after John Palliser, a British explorer), a region that was declared unfit for agriculture. The development of irrigation has made this region one of the most productive agricultural areas in Canada. Therefore, irrigation is a virtual necessity in the southern part of the province.

The application of supplementary water, also known as irrigation, started in the province in the late 1880s. By the early 1900s, major irrigation projects were operating. In 1914, the province of Alberta passed the Irrigation Districts Act. This provided the basis for irrigation water use under 13 irrigation districts. However, some private irrigation also exists in the province, although these areas are spread throughout the province. At the present time, 625,000 hectares are irrigated in the province. Irrigation in the province is a major user of water. Beaulieu et al. [3] estimated that in 2001, irrigation in Alberta used 2.9 billion m^3 of water, some 66% of total irrigation water in Canada. Thus, on average every hectare of irrigated land receives 5,524 m^3 of water annually.

The source of water for virtually all irrigation in Alberta is surface run-off. The majority of this water comes from snowmelt on the eastern slopes of the Rocky Mountains. Groundwater, although available, is not used for irrigation in any appreciable quantities.

1.2 Nature of technology used and crop mix

Irrigation in southern Alberta is effected using a system of more than 8,000 kilometres of conveyance works and more than 50 water storage reservoirs. Across the irrigated land in the province, irrigation producers are able to grow a great diversity of crops, many of which can be successfully produced only under irrigation, partly because naturally-available moisture for crop growth is generally less than half that required and is very erratic in its distribution. With more than 40 different types of crops being grown, irrigation water users have the flexibility needed to sustain viable farm enterprises, even during times with market swings in commodity prices.

1.3 Irrigation and diversification

Irrigation in Alberta is unquestionably a significant part of the agricultural landscape, occurring on less than 6% of the cultivated land base in the province, but contributing more than 19% of the gross primary agricultural production. In addition, irrigation provides avenues for diversification on farms as well of the regional economy. Irrigation is also an effective drought mitigation measure.

2 Need for the study

Water requirements for irrigation vary from year to year depending on weather conditions. In general, weather is the greatest factor driving demand for irrigation water, although the type of crops being grown also influences its demand. Demand for irrigation water is highest when it is a relatively hot and dry year, but those are also typically the years when the available supply of irrigation water tends to be less. In wetter years, there is generally more water available to divert but crops may not need as much irrigation water. This balance between supply and demand can be partially addressed by adoption of water conservation measures.

With irrigation, the majority of water that is applied to crops is taken up by the plants for growth, or evaporated. There is also some amount of water that is never used for irrigation itself, but is needed to maintain the minimum depth of water in canals and reservoirs in order to transport irrigation water through the system; therefore some of this water ends up as return flow back into other creeks and rivers.

As water supplies dwindle under the threat of climate change, several environmental lobby groups have argued that less water should be made available for irrigation (with the hope that more water would then be available for environmental objectives). The increasing criticism of water used for irrigating crops could threaten its long-term sustainability. The agricultural industry may be able to counteract criticisms of environmental lobby groups by improving irrigation water use efficiency. Irrigation water users and their organizations (including public bodies) are constantly evaluating their infrastructure and irrigation practices to minimize water losses (through seepage and evaporation) and maximize the area that can be irrigated. Improving water use efficiencies mean that less water is needed to produce the same amount of crop, or, that more area can be irrigated with the same amount of water. As available water is fully allocated, the only way to increase irrigation is through improving its efficiency of use.

One of the major attempts to bring the need for water conservation into public attention and focus was release of a *Water for Life* strategy that was released by the government of Alberta in 2003 [4]. The strategy includes measures to improve ability to capture and store water during high flow periods and to improve water use practices through conservation, efficiency, and productivity efforts, thereby tackling the challenge of unpredictable water supplies. Water conservation, efficiency, and productivity are fundamental in this water strategy.

3 Measurement of water use efficiency

Conserving water and improving efficiency of water use are somewhat related concepts. Improving water use efficiency is one of the measures that can lead to water conservation and thus, making water available for alternative users. In light of almost fully allocated water availability, improving water use efficiency is most welcome by water managers. A question arises as to what is an appropriate measure of water use efficiency.

Several researchers have used different definitions and concepts for measuring irrigation water use efficiency. Some define it as a dimensionless measure in terms of a ratio or percentage while others define it in terms of outputs measured in physical or monetary units per unit of water used. Variations and refinements in the definitions of the numerator and denominator also are common in the literature. For example, Seckler et al. [5] defined irrigation water use efficiency as the ratio of the net evapotranspiration to the amount of water withdrawn or diverted, where net evapotranspiration is the crop evapotranspiration requirements minus effective precipitation. Dhehibi et al. [6] presented a method of measuring efficiency of irrigation water use based on a non-radial notion of input-specific technical efficiency derived from a stochastic production frontier. This measure referred to the irrigation efficiency as the ratio of minimum feasible water use to the observed water use considering the output and inputs used with the production technology. Solomon and Burt [7] defined 'irrigation sagacity' as the ratio of irrigation water used 'beneficially or reasonably' to the irrigation water applied adjusted for the change in the storage of irrigation water. This definition is somewhat different from the authors' earlier definition of 'irrigation efficiency.' Obviously, these criteria require increasingly sophisticated and hard to obtain data. The terms 'beneficially or reasonably' in the definition of [7] are more likely to invite controversy rather than providing a more accurate measure of irrigation efficiency.

In contrast, the OECD [8] and Tollefson et al. [9] have defined water use efficiency in terms of physical and economic outputs per unit of water use, which could more easily be calculated with data commonly found in published sources and government documents. The first indicator of water use efficiency is called water use technical efficiency (WUTE) and defined as the mass of irrigated output divided by the volume of irrigation water utilized (e.g., tonnes/dam^3) while the second indicator is called water use economic efficiency (WUEE) and defined as the dollar value of irrigated output divided by the volume of irrigation water utilized (e.g., \$/dam^3). Even though the concept of efficiency is generally perceived in terms of a percentage or a unit free measure, the WUTE and WUEE, nonetheless, provide common sense yardsticks of the opportunity cost of a unit of irrigation water in physical and economic terms. Because of their computational ease, these two criteria of water use efficiency are evaluated in this paper in the following sections.

3.1 Water use technical efficiency

As mentioned earlier, the OECD [8] report defines water use technical efficiency (WUTE) as mass of agricultural production (in tonnes) per unit of water used. The WUTE is typically conceptualized in terms of level of irrigation water use and all the productivity is credited to it. The crop water use needs to be adjusted for return flow and precipitation amounts.

3.2 Water use economic efficiency

Technical water use efficiency indicator assumes that all crops grown under irrigation have the same impact on the well-being of the irrigator. However, such an assumption is not supported by real situations. Some irrigated crops are higher value crops while other crops are very low in value. Further complications arise from the fact that some crops under irrigated production systems are intermediate crops and therefore, not typically sold in a market place. A case in question is forages, which are primarily used for beef cattle enterprises. Water use economic efficiency (WUEE) removes this limitation of the technical water use indicator. It is defined as the ratio of total monetary value of agricultural production per unit of irrigation water utilized. Estimation of WUEE also is fairly data intensive since it requires knowledge of market prices, and disposition of various crops, in addition to cost of production. Further complications arise when some crops are marketed directly while others are for on-farm use and, therefore, marketed indirectly.

A major limitation of this indicator is that it is highly susceptible to changes in several factors, including choice of crops planted, agronomic/cultural practices employed, and crop prices. For example, adding more fertilizer to a crop usually increases its yield, and unless market price changes as a result, also increases the gross revenue obtained. This would result in a higher estimated value of WUEE but the change is contributed by factors other than improved water use technical efficiency.

4 Methods

In this study, the above two water use efficiency indicators (WUTE and WUEE) were estimated for irrigated agriculture in two river sub-basins – Oldman river sub-basin and the Bow river sub-basin, in Alberta following the definitions of OECD [8]. These sub-basins were selected as these have the largest area of irrigation in Alberta. The Oldman river sub-basin is about 75% larger than the Bow river sub-basin in terms of the volume of water diverted. The location of these two sub-basins is shown in Figure 1.

The total production (in metric tonnes) that was grown in each of these river sub-basins was calculated for each of the years 2004, 2005, 2006, 2007, and 2008 (see Yan et al. [10] for details). The total mass was calculated by multiplying the yield in metric tonnes by the number of hectares for each crop in each year. The yield was held constant at the 2000-2004 average to ensure that

Figure 1: Location of the Bow river sub-basin and Oldman river sub-basin in the South Saskatchewan River Basin (SSRB), Alberta.

the annual changes in yields would not affect the calculation of water use efficiencies. Since small changes occur annually in areas planted to each crop under irrigation (in response to output and input prices, rotational considerations and other factors), the total mass produced in each of the river sub-basins changes year-by-year.

Two measures of the value of production were calculated for each of the river sub-basins: gross revenues and net revenues (gross revenues less variable costs of production). Annual realized farm level prices were multiplied by average yields and by hectares of each crop to arrive at the gross revenues of crops produced. All variable costs of production were subtracted from the gross revenues to arrive at the net revenues.

The denominator for each of the efficiency measures is the gross water diverted to the irrigation districts in these two sub-basins in each year. For these calculations, we used the somewhat broader measurement of total water diverted from each sub-basin in dam^3 (AAFRD [11]). In addition to the water used for irrigation, this includes water used for industries, municipal, domestic, and environmental purposes. Due to the unavailability of data, return flows are also not subtracted from the gross diversion volumes. Thus, the data used in the denominator are slightly larger than the actual volume of water used for irrigation. However, the predominant use of water diversions in these two sub-basins is for irrigation and it is assumed that the water used for other purposes is constant from one year to the next.

All data for this analysis came from AAFRD [11–13], which reported areas planted to each crop, yields, prices, costs of production, and water diversion volumes during the time period studied.

5 Results

Table 1 presents data on the gross diversion of water in the Oldman and Bow river sub-basins of the SSRB for the period 2004-2008 while Table 2 presents calculations for the two measures of water use efficiency – WUTE and WUEE. The WUTE averaged 4.411 over the five-year period for the Oldman river sub-basin and 2.722 for the Bow river sub-basin. This means that a higher tonnage of crop is produced per dam^3 (1000 m^3) of water used in the Oldman river sub-basin. Water use technical efficiency also exhibited variability over time in each of the river sub-basins, perhaps due to a combination of factors that are external to the amount of irrigation water used. Examples of these external factors include planting of different crop mixes to take advantage of changing crop prices with different crops requiring different amounts of water, changes in the fertilizer mixture used to produce the crop resulting in changes in yield, differing soil composition between regions, and unexpected events such as weather, disease, and insects. Soil composition differences may also partially account for the difference in water use efficiency between the two sub-basins.

Table 1: Gross diversion of water 2004-2008 (1000 m^3).

Year	Bow river sub-basin	Oldman river sub-basin
2004	457,056	693,471
2005	375,183	596,919
2006	370,081	700,072
2007	442,589	897,693
2008	437,565	753,329
Average Std. dev.	416,495 40,715	728,297 110,217

Two measures of water use economic efficiency were calculated for the two sub-basins for the period of 2004-2008: WUEE (gross) and WUEE (net). The WUEE (gross) is the measure suggested by the OECD [8] and uses the gross revenue from production as the numerator. The WUEE (gross) averaged $289.58 per dam^3 over the five-year period in the Bow river sub-basin and averaged $423.25 per dam^3 in the Oldman river sub-basin (Table 3). The WUEE (net) averaged $114.56 per dam^3 in the Bow river sub-basin and $170.87 per dam^3 in the Oldman river sub-basin. The major advantage of the WUEE (net) calculation is that it accounts for changes in costs as well as revenues from crop production

Table 2: Water use technical efficiency (WUTE) 2004-2008 (tonnes/ 1000 m^3).

Year	Bow river sub-basin	Oldman river sub-basin
2004	2.557	4.520
2005	3.093	5.373
2006	3.134	4.690
2007	2.566	3.599
2008	2.262	3.872
Average Std. dev.	2.722 0.378	4.411 0.701

Table 3: Water use economic efficiency (WUEE) 2004-2008 (Can $/ 1000 m^3).

	WUEE (gross)		WUEE (net)	
Year	Bow	Oldman	Bow	Oldman
2004	233.65	405.57	79.20	125.41
2005	307.64	468.12	128.98	207.42
2006	288.74	406.38	109.03	176.72
2007	267.75	345.34	101.88	135.75
2008	350.12	490.83	153.71	209.07
Average Std. dev.	289.58 43.58	423.25 57.55	114.56 28.21	170.87 39.14

and thus reflects more closely the profitability signals to which farmers respond. Since output prices vary year-by-year, there is quite a lot of variability in the measures of WUEE (gross). Of course, input prices also vary, as do the amounts of inputs committed. But there is less year-to-year variability in the WUEE (net) measure.

6 Discussion

In this study water use efficiency in the two largest irrigation sub-basins in Alberta was examined using both technical and economic criteria. Although each can be conceptualized in different forms, in this study WUTE was defined as the total tonnage of agricultural products produced in the river sub-basins per unit of water diverted (measured in cubic decametres or dam^3). The economic efficiency took the technical efficiency one step further by weighting the physical quantities by their respective economic (social) values, still expressing it per unit of water. Two different measures of WUEE were developed: one calculated gross revenue per unit of water, the other calculated net revenue per unit of water.

Results suggest that WUTE does not vary as much from year to year as does either of the two WUEE measurements. The coefficient of variation for the WUTE is lower than that for either the gross or net WUEE.

For comparisons over a period of time, the WUTE indicator has some weaknesses, including:

- One, moisture to produce a crop comes from two sources: natural rainfall and supplementary irrigation water. In this measure, natural precipitation is ignored. Since some years have above normal rainfall, while others may be drought-like, the WUTE can produce misleading and even contradictory results. Producers generally adjust the level of supplementary irrigation water for various crops under different weather conditions.
- Two, the measure, in addition to not being comparable over time, also is not comparable over various regions on account of differences in crop mix, productivity of soils, weather patterns and other factors. Bulky crops, such as forages, generally are higher tonnage crops than are cereals or oilseed crops, therefore distorting the measure. Furthermore, the numerator in this measure does not take into account the total amount of biomass produced for all crops. For forages, the amount included is the total harvestable biomass, whereas only the weight of the seeds harvested is counted for grains and oilseeds.
- Three, this indicator uses the amount of water used in the denominator, which may also suffer from different definitions – water diverted, water applied by producers, water consumed, among others.

The first limitation of the WUTE can be removed by adopting an alternate measure of technical efficiency – incremental WUTE. In this measure, only incremental yield over the dryland production system would be included. This is based on the assumption that the natural precipitation will produce a yield equivalent to the dryland production system and the remaining increase in the yield under irrigation is due to supplementary irrigation. However, calculation of this measure requires data on dryland yields under equivalent conditions.

With respect to the denominator – amount of irrigation water used, refinements in its definition also could produce an improved measure of WUTE. For example, Tollefson et al. [9] has suggested two measures of this water: water diversion and water consumed. The latter is the difference between water diverted and water returned back to the source (perhaps at another location). The latter would yield a better measure of irrigation WUTE. Furthermore, water diversions also reflect water delivery efficiency to producers. Improvements in water delivery efficiencies (such as converting open canals to pipelines) could significantly improve both WUTE and WUEE.

The economic efficiency measure is an improvement over technical efficiency since it provides a measurement that is closer to the overall objective of irrigation farmers – improve their profitability and thereby overall social well-being. However, refinements in this indicator also can be suggested, in addition to those noted above. These include defining the numerator as the extra revenue obtained above what would have occurred under a dryland system. This is a more complicated calculation and would require yield and costs data for dryland production of various crops and weighting the values to reflect an aggregate WUEE.

References

[1] Shady, A. M., *Irrigation, drainage and flood control in Canada.* Ottawa: Canadian International Development Agency, 1964.

[2] Dale-Burnett, L., Palliser Triangle. *The encyclopedia of Saskatchewan.* Canadian Plains Research Center, University of Regina, Regina, p. 678, 2002.

[3] Beaulieu, M. S., Fric, C., and Soulard, F., *Estimation of water use in Canadian agriculture.* Statistics Canada, Ottawa. 40 pp. 2007.

[4] Government of Alberta. *Water for life action plan.* Edmonton. 28 pp. 2006.

[5] Seckler, D., Molden, D., & Sakthivadivel, R., The concept of efficiency in water-resources management and policy (chapter 3). *Water Productivity in Agriculture: Limits and Opportunities for Improvement*, ed. Kijne, W., Barker, R., & Molden, D., International Water Management Institute, Colombo, Sri Lanka, pp. 37-51, 2003.

[6] Dhehibi, B., Lachaal, L., Elloumi, M., & Messaoud, E.B., Measuring irrigation water efficiency with a stochastic production frontier: an application for Citrus producing farms in Tunisia. American Agricultural Economic Association Conference Proceedings, pp. 449-457, 2007.

[7] Solomon, K.H. & Burt, C.M., Irrigation sagacity: a measure of prudent water use. *Irrigation Science*, 18, pp.135-140, 1999.

[8] OECD – Organization for Economic Cooperation and Development. *Environmental indicators for agriculture – methods and results.* Volume 3. Paris. 2001.

[9] Tollefson, L., Harrington, J., & Hogg, T., Water quantity and use efficiency indicators for Canada. p. 3-11. *Agricultural Water Quality and Water Use: Developing Indicators for Policy Analysis*. Proceedings of an OECD Expert Meeting-Gyeongju, ed. Parris, K. & Jung, P.K. The Republic of Korea, October 2003.

[10] Yan, W., Klein, K. K., & Le Roy, D.G., Estimating the net benefits of irrigation water for cropping activities in irrigation districts in southern Alberta. SAREC Report 2010-1, Department of Economics, University of Lethbridge, 2010, 58 pages.

[11] Alberta Agriculture, Food and Rural Development (AAFRD). Alberta irrigation districts – Alberta irrigation information for the years 2004-2008. Irrigation Division, AAFRD, Lethbridge, Alberta. 2005-2009.

[12] Alberta Agriculture, Food and Rural Development (AAFRD). AgriProfit$ benchmarks for Alberta crop and forage producers. Edmonton. 2004-2007.
[13] Alberta Agriculture, Food and Rural Development (AAFRD). AgriProfit$ cropping alternatives. Edmonton. 2008.

Alberta's drive to use market-based instruments for ecosystem services provision

G. L. Kerr[1] & H. Bjornlund[2,3]
[1]*University of Calgary, Canada*
[2]*University of Lethbridge, Canada*
[3]*University of South Australia, Australia*

Abstract

Governments globally are struggling to reverse negative trends of decreased ecosystem function. Due to their nature as landscape drainage 'sinks', aquatic ecosystems are particularly in jeopardy. In response to some of the significant impacts on land and aquatic ecosystems, the province of Alberta (Canada) has recently introduced institutional, legislative and policy initiatives explicitly promoting ecosystem services (ES) production through market-based instruments (MBIs). This paper presents a brief account of the current state of aquatic ES knowledge and use in Alberta. Further, the paper explores the critical role that property rights play in MBIs, as a design criterion and applies this to the key legislation and policies in Alberta that currently enable MBIs. With few MBIs currently enabled, this paper evaluates the property rights regime for one MBI, the current water market in southern Alberta. The strengths and weaknesses of the property rights for water transfers provide some lessons for Alberta as it implements new MBIs for ES provision.

Keywords: ecosystem service, market-based instruments, property rights, water markets.

1 Introduction

Governments globally are struggling to reverse negative trends of decreased ecosystem function. Due to their nature as landscape drainage 'sinks', aquatic ecosystems are particularly in jeopardy. The Ecosystem Services (ES) approach has evolved from this struggle [1]. The complex nature of ES has promoted a number of governments to look beyond their current 'command and control'

WIT Transactions on Ecology and the Environment, Vol 134, © 2010 WIT Press
www.witpress.com, ISSN 1743-3541 (on-line)
doi:10.2495/SI100191

regulatory approach to environmental management toward the incorporation of market-based instruments (MBIs) [2, 3].

Within its legislation and policy context, the Government of Alberta (GoA), Canada, has shown strong support for using the ES approach to support aquatic ecosystem outcomes and for the use of market-based instruments (MBIs) to encourage the production of important ES. It is critical to understand both the context for ES in Alberta and the critical design criteria for MBIs to ensure that MBIs effectively modify the way humans interact with the environment so that the needed level of ES is produced.

Two main types of MBIs for ES exist; one provides a direct payment for ES provision by paying resource users to implement certain acts on their land. The second is based on markets in which willing sellers and willing buyers trade in licenses to e.g. catch fish or use water, or permits to e.g. emit pollutants into the air or water bodies. This paper will discuss both types in the context of Alberta policy and legislation but will deal in more detail with the important property rights issue associated with trading in licenses or permits, using water as an example.

2 Ecosystem services

The concept of ecosystem services (ES), as the benefits that humans derive from ecosystems, became popular among governments and stakeholders because it connects actions to protect and restore ecosystems with an improvement of human well-being [5, 6]. A key issue with the production of ES is that most demonstrate public good characteristics in that they are non excludable and non rival. A farmer, for instance, would have less incentive to produce these public goods when their livelihood depends on the private goods, such as crops and livestock.

The literature shows that increasingly, governments in the United States, Europe, Australia and Canada, are recognising a need to encourage the production of critical ES on both public and private land and to integrate the ES into environmental policy and legislation [6–8].

Alberta has recognized that there have been some significant impacts on land and aquatic ecosystems and has responded in turn with a number of key institutional, legislative and policy initiatives, including a recognition that the environment and society are interconnected. The vision statement in the Alberta Ministry of Environment business plan commits that "Alberta's environment sustains a high quality of life". The plan also recognizes that achieving this requires an understanding of the ties between nature and the environment and specifically highlights the importance of ecosystem services [9].

3 Market-based instrument design: property rights

Drawing again on the public goods nature of ES, MBIs do not typically develop for public goods such as pollination or climate regulation. Governments have had to develop clear incentives to ensure that critical ES are produced.

Traditionally governments have relied on regulation for environmental objectives, but recently MBIs have emerged as potentially a more effective tool.

With any tool development, the design of the tool is critical. No tool can be simply replicated in another jurisdiction as the context (spatial, temporal, scientific, political, social, etc) and policy and legislative history must be understood [2, 3, 10].

There is a great body of literature about the design criteria for MBIs, including specific and measureable outcomes, appropriate scale and fit within legislative and policy context, clear property rights, information disclosure, flexibility, monitoring and enforcement, and tool evaluation [6, 11, 12].

3.1 Property right and the nature of ecosystem services

The Government of Alberta (GoA) has explicitly included MBIs as instruments in a number of key documents that relate to environmental outcomes. It is apparent that there currently is a preference for, and acceptance of, the use of MBIs for environmental management [8, 13–16].

The choice of MBI will depend partially on the nature and security of the rights that government is prepared to provide to polluters or resource users. Understanding the nature of these rights is critical to the successful design and implementation of MBIs as the characteristics of the property rights will drive what types of tools that can be developed and conversely, if there is a desire for a new tool, it will drive how the property rights must be defined [17].

The role that property rights play with MBIs and the public goods nature of most ES are so critically important that it warrants further discussion [6, 17]. The key feature that distinguishes MBIs from other tools is the need to clearly define the rights and obligations of the parties involved in a transaction associated with the provision of ecosystem services. Within current natural resource management the issue of property rights, such as a permit to use resources like fish, or a right to emit pollutants, is quite clear. There is less clarity when the unit traded is, for example, a nutrient credit gained from a non-point polluter such as a land owner implementing a Best Management Practice for reducing the emission of nutrients into the river. When the landowner sells this credit to a point polluter, or the government because it improves ecosystem services, the new buyer owns the credit, but which obligations does the seller still have and what happens if the actions the seller took do not continue for the agreed duration? Uncertainty may result in failure to meet the environmental outcome. The following are the key elements of property rights [6, 17, 18]:

- Exclusivity - The permit or licence must belong to a specified party that has the exclusive ability to exercise the rights contained in the permit. There must be clear title to the permit. It is valued because it protects the right holder from interference from others that may want access to the specific right.
- Transferability - The permit or licence holder must have a relatively unfettered ability to dispose of the permit in any manner that suits their interests. This means being able to buy, sell or rent the right in accordance with changing market conditions. This is valued because it allows the right holder to choose the best use of his capital and time by transferring or selling the right.

- Duration - The time period over which the right holder expects to have control and benefit from the right. This is important for many resource users since considerable capital expenditures are required to utilize the resource and benefit from its use.
- Scarcity - For an individual or licence permit to be considered valuable there must be a limited number of permits available and/or it must have specific characteristics that make it unique. As long as parties can obtain additional rights from the government, there is little incentive to acquire existing rights.
- Security or quality of title - This answers the question of how vulnerable the right holder is to a challenge on the right. A set of legal rights reinforced by a registration system must secure the property right and must be enforceable against third parties and be enforced by the proper authorities. Lack of security produces uncertainty, which is an impediment.
- Flexibility- The flexibility of the right will determine how well the use of the resource can be adapted to changing conditions and what the right holder is enabled to do without consulting others. Limited flexibility can result in delays, costly procedures and losses for the right holder.

4 The Alberta context

Alberta has experienced intense economic and population growth over the past decade resulting in increased activity on the landscape. Alberta has recognised that this increased pressure combined with the increasingly complex nature of environmental issues, the current regulatory approach has severe limitations: in effectiveness and efficiency and the increasing costs for monitoring and compliance [19, 20].

The ES concept is evolving rapidly in Alberta. An ES assessment was undertaken in 2006 for land use planning in Southern Alberta and categorized twenty ecosystem services that then were assessed for their importance [8]. This ES assessment was included in the 2009 terms of reference for the South Saskatchewan Regional Plan [21]. The initial work by this group has listed some priority ES, including water supply; the emerging water market in southern Alberta will be evaluated in Section five.

4.1 Current state of aquatic ecosystem services in Alberta

Aquatic ecosystems are recognised as landscape 'sinks' into which the surrounding watershed activities drain. The primary human-induced impacts on rivers come from water withdrawals (allocations to divert), dams, and land cover changes [22]. While critical human activities like agriculture and irrigation provide society with food, community and economic contributions, the water allocation from the river to these activities have trade-offs. The depletion of river flows fundamentally alter aquatic ecosystems and the ability to provide such services as flood prevention; regulating runoff; water supply; assimilative capacity and improved quality of surface and ground waters; erosion and sediments control; bank stabilization; water infiltration in soil; groundwater recharge and cultural services, such as recreational, aesthetic and spiritual [1, 5].

Within Alberta, there are significant differences in the state of aquatic ecosystems and the way in which they are used across Alberta, especially north and south illustrated by the presence of irrigation districts in the south and drainage districts in the north. The region with the most severe impacts on its aquatic ecosystems is Southern Alberta. In 2003, the GoA undertook instream flow needs (IFN) studies in the South Saskatchewan River Basin (SSRB) to understand the aquatic ecosystem needs and found that for healthy aquatic ecosystems about 70-80% of natural flow was required. Around the same time a study for the GoA found that 30 of 33 main stem rivers in the SSRB ranged from 'moderately impacted' to degraded [24, 25].

The issues with aquatic ecosystems and the reduction in ES is clear; however, just because the GoA developed an IFN does not mean it can protect it. In context of water licences in Alberta, licence holders have a set of rights, established in law, on a first-in-time first-in-right basis. Licences were historically allocated to facilitate the settlement of western Canada and enable agricultural/irrigation use, not to protect aquatic ecosystems and ES [26]. A consequence of this system is that current consumptive licenses have the oldest priority while any new IFN licenses will be dated with that date and level of risk. Hence, during scarcity the IFN will have lower priority to any previous licences. Currently the SSRB is said to be over allocated; whereby the needs of the environment and the needs of the water licence holders can not all be met.

Alberta has met this challenge by developing legislation and policy. Four key pieces are presented based upon the potential impact they have on aquatic ecosystems and the use of MBIs to improve the provision of ES: i) the Water for Life Strategy (WFL) [27]; ii) the *Water Act* [28]; iii) *Alberta Land Stewardship Act* (ALSA) [29]; and iv) the Land-Use Framework (LUF) [30]. The following section will tease out the ES implications and will assess the MBIs included in these documents, looking specifically at the intended purpose, the current state of use and the clarity of property rights.

4.2 The *Water Act* and water for life strategy

The *Water Act* is the principal legislation governing water management in Alberta. Like many other jurisdictions the GoA owns the water on behalf of its citizens and is responsible for administering water legislation and regulations. Historically license holders were granted perpetual licenses to use water; while more recent licenses have been for a specified duration. The *Water Act* [28] relies heavily on regulation however it does specify that water transfers can be enabled through an approved Water Management Plan (WMP). To date only one of seven river basins in Alberta have an approved WMP, the South Saskatchewan River Basin (SSRB). In 2006 the approval of this WMP lead to both the closing of the basin to new surface water licences and the introduction of water transfers, leading to the birth of a water market.

The Water for Life Strategy (WFL) established three priority outcomes one of them being healthy aquatic ecosystems [27]. This was to focus the GoA and stakeholders on building strategies and tools to promote healthy aquatic ecosystems to secure the ES Albertans obtain from them. As such, WFL specifics three related objectives that pertain to ES and MBIs: i) develop a policy framework for ecosystem services markets in collaboration with Institute for Agriculture Forestry and Environment; ii) enhance the water rights transfer system; and iii) implement market-based ecosystem incentive program to meet water conservation and productivity objectives [27]. At present the only concrete action on these objectives has been an announcement by the GoA that they are undertaking a water allocation management system review.

4.3 Alberta land stewardship act and land-use framework

The 2008 Alberta Land Use Framework (LUF) and its resulting legislation the 2009 *Alberta Land Stewardship Act* (ALSA) are emerging as critical legislation and policy to connect aquatic ecosystems and land use actions and management [29, 30]. Both LUF and ALSA recognize that the environment provides "public goods" that Albertans enjoy and value. The GoA is currently trying to find new ways in which to share the costs of conserving them such as new policy instruments to encourage stewardship and conservation on private and public lands [31].

ALSA was made into legislation in 2009 and perhaps its most significant impact is that it has legal precedence over all other acts that deal with the environment and natural resource management. This power specifically states that if there are conflicts between ALSA and other acts, ALSA prevails. The enabling of MBIs and the promotion of ES may influence other Ministries and legislation in the future and drive environmental and natural resource management. The impacts this may have on current MBIs such as the water market in the SSRB remain to be seen.

LUF and ALSA heavily promote the use of MBIs; however there is a bit of confusion in that the MBIs presented in LUF and ALSA are not all the same. LUF lists tradable disturbance rights, transfer of development credits, land trusts, conservation easements, conservation offsets, and lease-swapping; while ALSA enables conservation easements, conservation directives, stewardship units, conservation off-sets, and transfer of development credits to encourage the conservation and stewardship on private as well as public lands. As ALSA is legislation it is likely that the tools in ALSA will be the focus on the GoA work for the next few years.

With the exception of conservation easements, none of the ALSA tools have been developed to date and the information within the LUF and ALSA is quite limited; however based on the information available the following paragraphs introduce the tools, the intended purpose, current status, and property rights implications.

- Stewardship Units - Stewardship units (SU) are not clearly defined in ALSA and do not appear in LUF at all. The purpose of SU would be to provide a unit of measure for stewardship, such as a hectare of a certain vegetation or wetland that would be the basis for a number of market-based incentive tools and could create a common property right that is to be exchanged across different land-use categories. It is clear that the SU do not, and may not constitute an interest in land, but would likely be an ecosystem services or bundle of ecosystem services [31]. This has significant implications or direction for MBI development, especially if all MBIs must use the SU.

- Conservation Easements - Conservation easements (CE) are a one time payment to landowners to preserve some ES. While non-government groups like Ducks Unlimited Canada area already using this tool, ALSA formalizes the government support. A landowner may grant a CE for any, or all of their land. A CE constitutes an interest in land that can be registered in the Land Titles Office that provides some security with respect to the right. Most existing conservation easements in Alberta have not been secured in this way and have therefore not been binding on subsequent owners. Consequently, some CEs have been dishonoured by subsequent landowners.
- Conservation Off-set Program - The creation of conservation offset programs is enabled; however ALSA does not elaborate when or how this tool might be used to compensate for impacted landscapes. LUF states that land conservation offsets are compensatory actions to address biodiversity or natural value loss (ecosystem services) arising from development on both public and private lands [31].
- Transfer of Development Credit - Transfer of development credits (TDC) allow for economic development on private lands, but direct it away from specific landscapes; however ALSA contains no description of the purpose behind a transfer of development credits scheme (TDC scheme). The TDC scheme will have to have the same purposes and principles as a CE.

In summary, ALSA and LUF provide a large number of MBIs that could positively impact ES production. At present only conservation easements are functioning. The lack of clarity and security of the property rights of CEs creates uncertainty for groups wanting to use them to secure easements. The evolution of the SU will be important to monitor. The development of a single type of unit of exchange for all ES will potentially make it easier to manage MBIs; however the practicality of being able to establish a single unit of measure for all ES will be exceptionally challenging given the nature of ES [4].

While Alberta is an apparent proponent of MBIs for environmental outcomes, there is limited experience with implementing these tools. The emerging water market is perhaps the best example to date of an MBI which can potentially be used for ES provision, through its ability to impact aquatic ecosystem health. The next section will explore the water market, assess the nature of the property rights and provide a few examples of how the aquatic ecosystem could benefit more directly from this market.

5 Alberta's emerging water market

Alberta Environment is mandated with the delivery and assurance of water allocation through the *Water Act,* including issuing licences and approving transfers. Within the SSRB there are approximately 13,000 water licenses. Similarly to many jurisdictions around the world agriculture and specifically irrigation is the largest user of water. All 13 of Alberta's Irrigation Districts (ID) are located in the SSRB and account for 75% of the total volume of water allocated, have the most senior (secure) licences and occupy 75% of the land [26, 32]. There has been increasing pressure from environmental groups in Alberta to ensure that water can be freed up from current licence uses to enhance aquatic ecosystem health [33].

The 1999 *Water Act* enabled water transfers with an Approved WMP. The first such plan was approved for the SSRB in 2006 and allowed for water transfers and closed three of the four sub basins to new surface water licences: a young quasi-market is born. To date transfer or trading activity has been very small and only 31 transfers have occurred [34].

While there are a number of reasons cited in the literature for why the market has so few trades even with the water scarcity issues and closed basin, this paper has focused on the role that property rights and the characteristics of those rights play in building certainty for licence holders to participate in a market, Table 1 highlights the strengths and weaknesses based upon the six characteristics listed above.

As noted in Table 1, water licence holders have usufruct rights to water in Alberta. Licence holders have exclusive right to use their licence based upon the conditions held in the licence and most licences were issues in perpetuity which gives added certainty in the right. There is also flexibility with regard to what the licence holder can do with the licence; however changes to the licence require government approval and some changes, like a transfer, are subject to appeal, which can increase transaction costs, opportunity cost and time.

The transferability of the licence is fairly clear at present, although it has high transaction costs, including a provision in the Water Act that allows for a 10% hold back on the transfer if deemed important for the aquatic ecosystem by the Director. In addition, irrigators in an ID have added 'red tape' as the licence is held by the ID on behalf of the irrigators within it and is subject to plebiscite votes and GoA approval for changes in licence.

There is a high level of certainty with most aspects of the water right. Perhaps the key challenge is the GoAs ability to cancel a licence under the emergency conditions in the Water Act or changes made as a result of the current water allocation system review.

Given the characteristics of property rights associated with water licenses in the SSRB and the emerging water market an obvious solution to enhance aquatic ecosystem health and the production of ES would be to transfer or buy water for the environment. There are a number of possibilities that have occurred in other water markets, including government purchase of licences, water trusts holding licences, private parties holding water licences for specific ecosystem functions or services (e.g. flushing the cottonwoods trees at critical times) and expropriation of water by government. Currently the *Water Act* only allows the government to hold licences for in-stream purposes, while Ducks Unlimited an NGO holds very large licenses for flooding wetlands, which is possible because the water is diverted for this purpose. However, the current Alberta water allocation system review is hearing from a number of parties that this should be modified [26, 36].

Table 1: Water licence property rights in the SSRB.

Property Right	Strengths	Weaknesses
Exclusivity	Water licences are registered exclusively to a licence holder for a specific purpose, diversion point, rate of flow, volume and priority date. The licence holders have usufruct rights.	The Water Act gives the GoA the emergency authority to suspend a licence [27, Part 7. 107]
Transferability	The approved SSRB Water Management Plan enabled transfers of all or part of a licence that is in 'good standing'	Transfers are subject to the Water Act and the SSRB WMP and can have no adverse impact on the aquatic environment or third parties [21]. Irrigators face more restrictions within an Irrigation District (ID) as they can only transfer inside the ID. Transfer process has high transaction costs, is complicated and time consuming [35].
Duration	Most licences were granted in perpetuity. Some licences have specific terms and conditions that restrict withdrawals at below specified river flows giving priority to environmental requirements.	Threat of government modifying existing licences as a result of the current water allocation system review is a concern. Since 1990 licenses have been granted a 10 year renewable licence
Scarcity	The SSRB is closed to new water licences; if anyone wants water they must buy a licence. In the SSRB water scarcity is a constant issue or threat on the horizon	Most current license holders have excessive allocations relative to their needs and hence can manage quite severe scarcity
Security of title	Existing water licences provide clearly defined usufruct rights that are legally enforced. Water licences are registered exclusively to a licence holder for a specific purpose, diversion point, rate of flow, volume amount and priority date.	Have to be in good standing for three years or entire licence can be cancelled. Water metering and actually mandatory reporting are not complete which may detract from real or perceive security of title

Table 1: Continued.

Property Right	Strengths	Weaknesses
Flexibility	There are a number of 'products' that can be traded - permanent and temporary transfers of whole or part of license or assignments can be made of whole or part of seasonal allocations.	Any modification of a licence has to be approved by the government. Some changes are minor while others may require public consultation and a more time consuming process. There is less flexibility in an ID as a plebiscite with 50% is necessary for major changes like land expansion or transfers [21].

6 Conclusions

Governments are struggling to reverse negative trends of decreased ecosystem functions and aquatic ecosystems are particularly in jeopardy due to their nature as landscape drainage 'sinks'. Within Alberta the Ecosystem Services (ES) approach has been incorporated into a number of policies and legislation to promote the health of aquatic ecosystems. MBIs are promoted as an appropriate tool for managing ES, but must be designed carefully given the public goods nature of most ES. The characteristics of the property rights associated with the units, such as the Stewardship United discussed in this paper, that are traded and which defines how the ES will be provided, are critical to the design and successful implementation of the MBI and to the successful provision of the ES. If a landholder or right holder does not have confidence or clarity about what their rights are and what they are expected to do, there may be a resistance to adopting MBIs.

While Alberta has explicitly promoted the use of MBIs for production of ES and introduced MBIs as part of resent legislation, these policies and pieces of legislation are quite new. There is little evidence of implementation of these new tools. The next few years will be a very interesting time to watch Alberta to see what will happen with the numerous MBIs that have been enabled. The nature of the property right associated with the provision of ES will be a critical consideration in the design of any of the potential MBIs that have been enabled.

References

[1] Millennium Ecosystem Assessment, *Ecosystems and Human Well-being: Wetlands and Water Synthesis*. Millennium Ecosystem Assessment Series. Island Press Publications: Washington, D.C, 2005.

[2] de Loë, R.C., Armitage, D., Plummer, R., Davidson, S. & Moraru, L., *From Government to Governance: A State-of-the-Art Review of Environmental Governance. Final Report.* Prepared for Alberta Environment,

Environmental Stewardship. de Loë Consulting Services Guelph, ON, 2009.

[3] Coggan, A., Whitten, S.M., and Langston, 2005. *Nesting MBIs in current institutions and structures – can it be done and what are the implications?* CSIRO Sustainable Ecosystems, 2005.

[4] Stoneham, G., An overview of the ideas and information needed to develop and implement MBIs. Proceedings of the 6h Annual AARES National Symposium. Canberra, 2004.

[5] Ranganthan, L, Raudsepp-Hearne, Lucas, N., Irwin, F., Zurek, M., Bennett, K., Neville, A. and West, P., *Ecosystem Services: A Guide for Decision Makers.* World Resources Institute, 2007.

[6] Landscape Change Team, *Ecosystem Services through Land Stewardship Practices: Issues and Options.* Land Management Group. Department of Sustainability and Environment. Victoria Australia, 2003.

[7] Mistakis Institute. *Alberta Ecological Goods and Services Scan and Recommendations to NAWMP* Prepared for Alberta North American Waterfowl Management Plan (NAWMP), 2008.

[8] Integrated Environments and O2 Planning and Design (Integrated), *Ecosystem Goods and Services Assessment – Southern Alberta, Phase 2 Report.* Publication: Alberta Environment, 2007.

[9] Alberta Finance, Alberta Environment Business Plan 2009-12. Online. http://www.finance.alberta.ca/publications/budget/budget2009/envir.pdf

[10] United Nations Environment Program, *The Economics of Ecosystems and Biodiversity (TEEB) for National and International Policy makers*. 2009.

[11] Kamarck, E.C., *The End of Government As We Know It: Making Public Policy Work.* Lynne Rienner, Boulder, CO. 2007.

[12] Chaudhri, V. *Market Based Instruments and NRM:* Proceedings of the 6h Annual AARES National Symposium. Canberra, 2004.

[13] Institute for Agriculture, Forestry and the Environment, *Market-based Instruments. Fact Sheet.* Alberta Agriculture, Alberta, 2008.

[14] Integrated Environments, *Ecosystem Goods and Services Assessment – Southern Alberta, Phase 2 Report: Conceptual Linkages and Initial Assessment.* Published by Alberta Environment, Alberta, pp. 131, 2007.

[15] Alberta Water Council. Alberta Water for Life Strategy, Alberta Online. http://www.waterforlife.gov.ab.ca

[16] Alberta Environment. Environmental Assurance Division. Environmental Tools Guide. Online. http://environment.alberta.ca/1985.html)

[17] Alberta Government. Land-use Framework. Online. www.landuse.alberta.ca/documents/Final_Land_use_Framework.pdf

[18] Scott, A., Introducing Property Rights in Fishery Management (Chapter 1). *Use of Property Rights in Fisheries Management. Fisheries Technical Papers*, ed. R. Shotton. Fisheries and Agricultural Organization of the United Nations, pp.1-13., 1999.

[19] Bjornlund, H., Efficient water market mechanisms to cope with water scarcity. *International Journal of Water Resources Development.* **19(4),** pp.553–567. 2003.

[20] Wenig, M.M., A.J. Kwasniak, & Quinn, M.S., Water Under the Bridge? The Role of Instream Flow Needs (IFNs) in Federal and Inter-Jurisdictional Management of Alberta's Rivers. Proceedings of the Alberta Society of Professional Biologists, ed. Alberta Society of Professional Biologists, Edmonton, Alberta. 2006.

[21] Bankes, N., The Legal Framework for Acquiring Water Entitlements from Existing Users. *Alberta Legal Review,* **44(323)**, 2006.

[22] Alberta Government (AG), 2009 Terms of Reference for Developing the South Saskatchewan Region http://www.landuse.alberta.ca

[23] Palmer, M.A. Lettenmaier, D.P., Poff N.L., Postel, S.L. Richter, B. & Warner, R., Climate Change and River Ecosystems: Protection and Adaptation Options. *Environmental Management.* **44**, pp.1053–1068, 2009.

[24] Alberta Environment, Approved Water Management Plan for the South Saskatchewan River Basin in Alberta. Edmonton: Alberta Environment, 2006.

[25] Alberta Environment, *Draft South Saskatchewan River Basin Water Allocation.* Online www3.gov.ab.ca/env/water/regions/ssrb/studies.asp

[26] Kwasniak, A.J., Quenching Instream Thirst: A Role for Water Trusts in the Prairie Provinces. *Journal of Environmental Law and Practice* **16(3)** pp. 211, 2006.

[27] Alberta Environment, Alberta Water for Life: Strategy for Sustainability. Online http://www.waterforlife.gov.ab.ca/docs/strategyNov03.pdf

[28] Alberta Government. Alberta Water Act. Online. http://www.canlii.org/en/ab/laws/stat/rsa-2000-c-w-3/latest/rsa-2000-c-w-3.html

[29] Alberta Government. Alberta Land Stewardship Act (ALSA). Online http://www.landuse.alberta.ca/AlbertaLandStewardshipAct/Default.aspx

[30] Alberta Government. Alberta Land-use Framework. Online. http://www.landuse.alberta.ca/AboutLanduseFramework/Default.aspx

[31] Poschwatta-Yearsley, J. & Zelmer, A., The Alberta Land Stewardship Act: Certainty or Uncertainty? *Canadian Institute for Resource Law.* University of Calgary. **106**, 2009.

[32] Alberta Environment, South Saskatchewan River Basin Planning Program: Summary Report. Government of Alberta, 1984.

[33] Droitsch, D. & Robinson, B., Share the Water: Building a Secure Future for Alberta. Water Matters and Ecojustice. pp. 5-33, 2009.

[34] McGee, D. Personal communication, 5 January 2010, Director, Southern Saskatchewan River Basin Water Management Plan Implementation, Lethbridge, Alberta.

[35] Nicol, L., Klein, K. & Bjornlund, H., Permanent Transfers of Water Rights: A Study of the Southern Alberta Market. *Prairie Forum,* **33(2),** pp.341-56, 2008.

[36] Alberta Water Council. Water Allocation Transfer System Upgrade. Online. http://www.albertawatercouncil.ca/Portals/0/pdfs/WATSUP_web_FINAL.pdf

Trade influences in Australia's allocation market: can allocations provide environmental water?

A. Loch[1] & H. Bjornlund[1,2]
[1]*Centre for Regulation & Market Analysis, School of Commerce, University of South Australia, Australia*
[2]*University of Lethbridge, Canada*

Abstract

Over the last 25 years water markets in Australia have been used to transfer water entitlements and allocations between irrigators, helping them to manage the risks associated with water supply and the demand for structural change. More recently governments have entered entitlement markets to purchase water for the environment. This paper hypothesises that a focus on purchasing entitlements will not motivate a sufficient number of irrigators to engage in the provision of environmental water. It is hypothesised that governments may need to enter the allocation market to secure long-term arrangements for counter-cyclical provision of environmental water. Avoiding any negative impacts of interference on the traditional users of this market requires a greater understanding of how irrigators currently use allocations, and what drives their decisions to buy and sell. This paper provides the findings of qualitative research involving 39 irrigators in three states within the Murray-Darling Basin in Australia. The findings provide a preliminary assessment of the factors influencing irrigators' decisions to trade allocations and how these factors change across and within seasons. Irrigator attitudes toward government use of the allocation market to support counter-cyclic environmental water provision are also examined. Overall, irrigators appear comfortable with the notion of trading in allocations with the government, rather than in entitlements alone.

Keywords: allocation trading, environmental water provision, trade influences, Murray-Darling Basin.

WIT Transactions on Ecology and the Environment, Vol 134, © 2010 WIT Press
www.witpress.com, ISSN 1743-3541 (on-line)
doi:10.2495/SI100201

1 Introduction

Many of Australia's rivers have been recognised as being over-allocated, with little water remaining for the environment and eco-system health. One such river system is the Murray-Darling Basin (MDB) in Australia's south-east. To address this situation governments at the state and federal levels have introduced policies to return water to the environment [1]. Initially emphasis was placed on the use of regulatory planning processes. However these processes have largely failed due to a prolonged period of extreme drought. More recently governments have changed their strategy and are now trying to achieve environmental outcomes by purchasing water entitlements in the water market.

Entitlements represent an access right to water. Without that access irrigation is impossible. To irrigators entitlements therefore constitute the single biggest asset they own, as they provide a degree of income certainty. In recognition of this irrigators have been reluctant to sell their entitlements. There is, however, an alternative water market in which the seasonal right to use water is traded rather than outright transfer of ownership—the allocation market. This market has been far more readily accepted by irrigators [2].

The premise of this research is that irrigators will be reluctant to sell their entitlements to governments. This research also hypothesises that the allocation market is more appealing to irrigators for the provision of environmental water. Warnings of dire market consequences from government intervention into the entitlement market have already been sounded [3]. Similar concerns would no doubt surface should the allocation market also be targeted for recovery purposes. A better understanding of how irrigators use this market, what influences their decisions and how these factors vary within and between seasons will allow the government to plan their environmental purchases. Appropriate planning based on such knowledge will cause minimum disruption to traditional market operations and thereby minimize socioeconomic impacts.

Previous research into water markets has examined the drivers of allocation price and volume traded, particularly in the Goulburn district of Victoria [4–6]. However, specific comprehension of what drives allocation trade decisions across and within seasons requires further investigation [6]. The objective for this research is to extend the understanding of allocation trade behaviour using cross-sectional analysis of water trade influences among irrigators in the southern MDB. The principle methodological approach is qualitative with a view to inform later quantitative research and more generalisable results.

2 Methods

2.1 Research design

This research began with the assumption that irrigators would be reluctant or unwilling to part with their entitlements, suggesting that alternative trade products would be needed to recover sufficient environmental water to meet government objectives. Alternative methods have been suggested in previous papers [7–9] but the government's potential entrance into the allocation market has never been tested amongst irrigators to gauge their attitudes. The intent of this research is to provide emergent theory or to extend the theoretical understanding [10] of issues related to irrigator trade in the allocation market, including attitudes to possible government intervention.

Using the interpretivist paradigm [11] the research design centred upon collecting qualitative data from irrigators who are actively trading in water markets. To broaden existing water trade knowledge irrigators in the three southern MDB states (i.e. New South Wales (NSW), Victoria and South Australia (SA)) were targeted. The collection of data followed a naturalistic methodological approach [12] into a phenomenon through the subjective accounts of those that have experienced it [13]. The experiential questions for the research were determined *a priori* in an effort to keep subjective discussions on a broadly focused track. These questions included:

- what influences irrigator trade in water allocations across and within seasons;
- what drives irrigator trade in entitlements; and
- would allocation products (e.g. option contracts or leaseback agreements) increase the willingness to sell water to the environment.

2.2 Sample

A maximum variation approach was used to identify the sample [12] so that common patterns could be identified in irrigators' use of water markets. The sample was selectively chosen [11] to represent small family to large corporate operators, district irrigators and private diverters, different irrigation zones and commodity groups, and a mixture of young and old irrigators.

The use of maximum variation sampling provided sufficient breadth within the sample, a short data collection timeframe minimised data distortions introduced by changes over time, and a minimum of one and a half hours for each interview provided depth of data [14]. The use of maximum variation and selective sampling techniques facilitate the collection of common and divergent experience patterns during a particular period. In this case, the experiences occurred during a period of prolonged drought, severe water scarcity, and low commodity prices. Irrigators in SA were hit particularly hard. Prior to 2006 they had never expected or experienced lower than 100% allocations. Further, due to the dominance of permanent plantings SA irrigators are very vulnerable to low allocations. Since 2006 SA allocations have not reached 100%, and irrigators there have experienced allocations as low as 18%. These contextual influences must be taken into account when analysing the data [15]. All participants were male, aged in their early 50s (average age 53 years) and >75% in each state were full-time irrigators earning a majority of their income from irrigated farming.

2.3 Procedure

Lists of potential participants were provided by irrigator organisations in SA and NSW, and by a commercial list provider in Victoria. Irrigators were contacted by telephone and assessed against the selected criteria. If suitable, the research project was described and where the irrigator indicated a willingness to participate an interview time was arranged. Eventually 39 candidates were chosen—14 in SA, 12 in NSW and 13 in Victoria. Prior to the interview each participant received a small questionnaire to be completed either prior to or during the interview. The mixture of qualitative and quantitative methods provides insight that each method alone might not achieve [16].

Following the tenets of interpretivist and naturalistic methodological design [12] a semi-structured interview was prepared, providing a schedule of questions to be explored with each irrigator. This approach avoided predetermined responses but ensured that basically the same information was obtained from each participant [15]. An interview guide was also prepared, leaving the interviewer free to explore or probe within and outside these predetermined areas of inquiry. During the interviews an approach of empathetic neutrality was sought to achieve a non-judgemental, balanced report [14, 15].

Interviews were conducted by the first author who had previous experience as an irrigation farmer. Hoepfl [15] suggests that shared background affords the interviewer with sensitivity to emergent theory and a capacity to gauge interesting lines of inquiry in the field. Shared experience also provides the interviewer the ability to process data as it becomes available, provide immediate feedback, seek further clarification, incorporate emergent issues from early into later interviews and explore atypical or unexpected responses [15]. The interviews averaged one-and-a-half hours, were audio recorded and transcribed. Interviews were typically conducted in the irrigator's home or place of business, although some were held in public places for convenience.

2.4 Analysis

This research is classified as grounded theory seeking [17]. The emphasis is upon further investigation as a consequence of the theory generated. In line with this approach, the constant comparative method was utilised to allow for the creation of generalised theoretical notions about the phenomenon under study, with the intent of further quantitative analysis of the theory at a later stage [18]. This process required the repeated comparison of text sections to identify patterns in farmer experiences. In this way theoretical foundations with explanatory power were eventually derived, as detailed below.

In line with the approach advocated by Aronson [19] the transcripts were firstly read to identify relevance to the issues of interest (e.g. within season allocation trade) as well as other emergent categories (e.g. the provision of ecosystem services). Similar concepts that described abstract representations of events or actions/interactions relevant to the research were open coded into categories. Each category collectively held examples of more than one related concept from the data. This process systematically summarised the data by relating criteria around a central issue [20]. Using NVivo software the categorised sections of text were next coded axially. This process identified underlying commonalities and attributes on which phenomena could be further categorised. The resulting analysis was then placed into a matrix that included main and sub-category listings, explanatory information about that grouping and possible research questions. It should be noted that the results are derived then from irrigator statements which may, or may not, reflect accurate facts.

Triangulation was also applied to the data as a means of generating trustworthiness in the analysis [14]. 'Methods Triangulation', as applied herein, is based on the degree of convergence present between the findings of this research and the work of prior quantitative studies. The closeness of convergence between findings on one hand, and the alternative theories offered by this research on the other hand lends credibility to the research and suggests further investigation. Finally, the emergent categories were compared to concepts or explanatory models presented in the existing literature.

3 Results

3.1 Across season allocation trade influences

Irrigators state that they tend to sell allocations on the basis of *income generation, covering fixed access charges, surviving until sale and exit or having surplus water.* A majority of irrigators who discussed the issue in the present scarce supply context reported that they were able to generate more income from water trade than farm production. That is, if water earned less from use in crop production than on the market, it would be sold seasonally for cash flow. Another group of irrigators reported that as long as the trade covered any fixed costs of water access they would be satisfied. These irrigators derived their income from farming, but recognised the burden of fixed charges in the current low allocation context and saw allocation trade as one way to meet such expenses. A small group of irrigators discussed allocation trade across seasons as a means of surviving until they could sell the farm and exit from the industry. These irrigators had sold parts of their entitlements to reduce debt or financial pressure, and were now awaiting an opportune time to sell remaining assets. Despite the drought a few irrigators reported having surplus water to trade as a consequence of efficiency gains. They now traded this water each year to supplement their income or to maintain a history of use.

Irrigators tended to buy allocations on the basis of *necessity and the current unusual allocation conditions.* Irrigators discussed purchasing across seasons to protect permanent crops or their existing investments in stock and equipment. Usually this investment had been based on historical access to reliably high allocation levels. Such activity was reported predominantly by irrigators in the horticulture industry. Very few dairy irrigators discussed buying out of necessity during the interviews, instead following a current strategy of stock reduction. Another group of irrigators discussed purchasing allocations because of the unusual situation at present, their limited experience with low allocations and the current uncertainty associated with high security licenses. Typically in the past, these irrigators had not used the allocation market as traditional entitlements were adequate. Now however, they were being forced to buy allocations to meet shortfalls in supply. In SA particularly, under the current drought, high security entitlements could not offer anywhere near the allocation level of similar high security entitlements in NSW, prompting the purchase of allocations in recent years by SA irrigators to supplement their dwindling allocations.

3.2 Within season allocation trade influences

3.2.1 Early season trade influences (Sep-Nov)

When discussing influences to sell allocations early in a season the reasons typically revolved around *price, nature of the entitlement* and *market distortions*. The issue of seeking to maximise profit from the early sale of allocations is not overtly discussed, but is clearly a driver for irrigators who can access their allocation early and trade while the market retains some volatility. Such volatility leads to higher early prices, which simplifies the decision for irrigators with annual crops. This is particularly the case in NSW where sizable allocations can be announced in the opening months of the season. Those with high security entitlements therefore have the opportunity to engage in profitable early trade. Finally, the prospect of market distorting activity such as trade embargos to limit the loss of too much water from any one area or the prospect of rule changes associated with trading can also drive irrigators to sell early in the season and avoid being excluded later. For example, an embargo on allocation trade occurred in the Murrumbidgee Irrigation District (MID) in 2009 and as such is a purely contextual driver of trade. Changes to early season carryover rules in the MID were also an example of uncertainty or market frustration for irrigators.

The decision to buy allocations early in a season is reportedly based upon issues of *cash flow, risk management* or *market interference.* If water is required to start or support an existing crop, and therefore needs to be purchased, one group of irrigators reported issues of finance as influencing their decision. Some need to access debt funding to support the crop immediately while others use recent commodity payments as a source of finance. If neither is available they cannot trade; or are forced to trade during the season as and when cash flow becomes available. Another group of irrigators discussed their desire to avoid the impacts of continued low allocation volumes during the season by purchasing water early as a risk management strategy. They seemed willing to pay higher prices to achieve this outcome—in a sense like paying insurance to avoid risk—especially if they had not managed to buy carryover water in the previous season or if carryover announcements varied between trading periods. Lastly, during the 2008 and 2009 seasons in SA the government announced offers of critical allocation water to assist horticultural growers to protect their valuable crops. The prospect of such water lessens the need to 'rush into' the market and allows some irrigators the option of waiting and seeing what the season, and state government does before committing to a decision. This is a good example of allocation market interference.

3.2.2 Middle season trade influences (Dec-Feb)

While irrigators who had not sold allocations early in the season due to risk aversion or indecision discussed accepting whatever sale price they could in the middle months, the majority of the issues in middle allocation trade appear to revolve around buying rather than selling. These issues include *price stabilisation, allocation* or *market predictions, seasonal variation* and *market distortions*. Irrigators in this group reported that once the early trade volatility and seasonal allocation uncertainty begin to settle, market prices stabilise. Those irrigators who are more accepting of risk might choose to buy now with the expectation of a lower price. Many irrigators discussed having learnt from panic buying in 2007, when allocations were expected to stay around 16% for the year and prices quickly escalated to $1,200/ML. Such panic buying is unlikely to be seen again and irrigators across the states indicated that they had set upper purchasing bounds of around $300-350/ML.

These same irrigators therefore study the market and seasonal allocations, using various sources of information. They attempt to enter the market at the 'best' time and price points. However, other factors such as seasonal variability and hot/dry periods during the spring and summer months may force higher than expected water use, in turn prompting middle season trade to secure additional water. Finally, as occurred in 2009, market distorting factors such as the federal government's refusal to execute entitlement purchase contracts previously promised to irrigators saw some Victorian irrigators enter the market unexpectedly during Oct/Nov of that season. These irrigators needed water to either irrigate new 'late' crops or improve the quality of existing crops since they once more represented the major farm income source.

3.2.3 Late season trade influences (Mar-May)

Irrigators who buy allocations late in the season typically discuss *balancing their water account, risk attitude, carryover demand* and *seasonal variability* as issues. For example, if irrigators have exceeded their usage during the year they will need to buy water to balance the account with their water provider to avoid hefty fines for excess use. This could occur in the middle of the season, but most irrigators prefer to do this at the end of the season, possibly because prices traditionally have been lower at that time. Alternatively the risk profile of the irrigator also influences their decision to buy late. If the irrigator is a risk taker, they discuss gambling on later water price reductions, preferring to see the season through and buying water for crop-finishing purposes. If the irrigator is more risk adverse, they report entering the market later to buy carryover water in preparation for anticipated low allocation announcements next year. In addition, a small group of irrigators had experienced longer than typical hot or dry periods in recent years, prompting their forced entry into the market later in the season to secure additional water.

When selling late in the season irrigators predominantly reported their influences for doing so as *expecting higher prices from late carryover demand, having surplus water to sell* and *avoiding the loss of unused water*. As mentioned above, if future water is expected to be scarce carryover demand will be high. Some irrigators with surplus water count on that for late season price 'kicks' and sell it at this time. A smaller number of irrigators also discussed losing unused water due to carryover restrictions and 'parking' that water on another irrigator's account with surplus carryover capacity. Once the new water year started, the restricted irrigator could access that water for opening supply. In these cases irrigators tried to beat what were seen as frustrating system rules.

3.3 Entitlement trade influences

Discussion on the topic of entitlement trade provided emergent sub-categories related to buying, selling and outright refusal to trade. When discussing buying entitlements a group of irrigators suggested that the *uneconomic nature of farming* and *high prices of entitlements* precluded purchase at present. Current poor returns from almost all traditional commodities mean irrigators cannot justify going into debt to finance the purchase of entitlements. Government demand for entitlements is also providing a price floor of around $2,000/ML for high security and $1,000/ML for general security entitlements, which irrigators can't justify or match in the current context.

On the topic of selling entitlements, the discussion centred upon *forced* rather than *willing sale, capitalisation to continue farming, retirement/resentment, future uncertainty or selling for the 'right' price*. When discussing the government program to buy water for the environment most irrigators stated that there were very few willing sellers, and that irrigators taking part had been forced to sell due to debt or other financial pressures. Another group of irrigators indicated that, while not pleased about selling, they had been willing to do so, capitalising their assets to cover debt or the cost of restructuring, while also continuing to irrigate with allocation purchases. Some irrigators reported feeling resentment toward or uncertainty about the future of irrigation. Such feelings or the prospect of eminent retirement with no family successor had influenced their decision to sell. Lastly, a few irrigators said that they were happy to sell their entitlements because they perceived market value would drop in future and now was the opportune time to sell. Others stated that they 'would' be happy to sell if offered the right price—typically well above current market offers. Additionally, there were some irrigators who, when asked about entitlement trade, simply commented that they would refuse to sell for any reason. These irrigators mentioned that they simply *couldn't or wouldn't sell their entitlements, or that community impact concerns prevented the sale*. Irrigators in this group stated that they need the entitlement and were using it for farming. They described the water as an asset and major part of their farm value and would only sell if that value was recognised; again typically well above market prices. A very small group of irrigators, predominantly in Victoria, discussed community impacts as the reason for not selling their entitlement.

3.4 Option/leaseback attitudes

A range of positive and negative attitudes emerged from discussion about the use of options or leaseback products to secure environmental water. The positive focused on *broader trade products, leaseback appeal, counter-cyclical funding opportunities* and *surplus water trade*. Almost every irrigator who discussed this topic expressed comfort about using allocation products to provide environmental water, and had no difficulty entering into agreements with the government. Discussion by some irrigators showed that they perceive the current entitlement purchasing policy as lacking in vision or short-sighted, and that an expanded list of trade products is warranted. The concept appealed more to annual croppers, especially upon the return to normal seasonal supply conditions. One group of irrigators picked up on the capacity of such programs to be self-funding through counter-cyclical trade arrangements (i.e. buying when allocations are high and prices low; selling when allocations are low and prices high), particularly if surplus water again becomes available and where fixed irrigator costs of access need to be met.

The negative discussion associated with this category included concerns about *previous bad lease experiences, efficient water delivery* and *pricing*. A few irrigators alluded to previously having bad experiences in long term leases with other farmers, especially in scarce supply years. Farmers need the water each year regardless, whereas governments might not. Another small group held concerns about how such water would be delivered through the system and that current environmental managers would be a barrier to such agreements. Finally, the issue of how such leases would be priced concerned a few irrigators, but if the price was acceptable the concept seemed worthy of further consideration.

4 Discussion

The results of this research offer valuable insight into the questions posed. In regard to the question of what influences allocation trade across and within seasons the prepared list of possible issues based on the literature proved useful, with expected issues appearing regularly through the discussions. However, other trade drivers also emerged. For example, previous research into irrigator activity in water markets mainly focused on price and volume traded, particularly in Victoria's Goulburn area. They did not examine three states at once but there are some interesting parallels and divergences. Bjornlund [21] discusses irrigators using the allocation market at different times to supplement or fully generate farm income, selling in some years and buying in others. As shown here these influences remain important factors in trade decision making, especially in the context of low allocations and during early season periods. Similarly Wheeler et al. [6] find that rainfall, evaporation and market conditions influence general allocation trade across seasons, but that to determine the real drivers of individual buying and selling would require more specific data. This study attempts to provide such data. Wheeler et al also found that as markets mature outlandish bids decrease as a function of irrigator experience, which is consistent with our findings on future panic trade. Seasonal effects on allocation trade, especially as a driver for late season activity based on lower prices have been reported in other studies [4], but drivers such as carryover based trade are new to the literature, having only been possible in the past few years.

The need to protect investment in permanent crops and capital equipment has also been found in previous research [5], particularly in the horticulture and dairy industries. A difference in this study is that dairy farmers no longer seem willing to pay premiums, and it is uncertain how long grape growers will continue purchasing with current poor commodity prices. Risk management strategies within and between seasons [22] and uncertainty have been linked to buying behaviour previously [23]. In this study the influence of risk attitudes seem particularly important for early or late within season decisions. It is not surprising that the current period of extended drought has caused farmers to place increased importance on risk management.

That irrigators utilise different allocation trade strategies across and within seasons in response to commodity prices, water prices, seasonal allocations, weather patterns and risk management strategies [21] is also supported by this research. However, other factors such as market distortions, fixed charges, surviving to exit, avoiding carryover losses and financial capacity to trade should perhaps also be taken into account. Further, the issue of end of season account balancing has previously been suggested as a late season allocation trade influence [6] and thus represents another variable of interest in the continuing research. Finally, another study by Wheeler et al. [24] found that the highest trade volume took place in January to March each year (middle seasonal trade). That is consistent with our findings of price and market stability during middle season, along with greater allocation certainty. Interestingly, Wheeler et al. suggested that the government can influence market activity by their involvement; this research has identified some specific examples of that interference.

The second question regarding influences of entitlement trade also shares some similar and divergent findings. For instance, that price is not yet a factor in entitlement demand [23] is a finding we would dispute in the present context. Utilisation of entitlement trade to, among other things, become debt free and stay in the community [25] is discussed in this study, but contrary to previous research there is little mention of lifestyle or off-farm income as drivers in the adjustment process. Equally, while the issue of community barriers to entitlement sales has featured in earlier studies [26, 27], particularly in public opinion or debate, it does not seem to share as much priority here in private reflections on the issue. Finally, some of the uses of entitlement sales identified in this paper—i) to delay exit for as long as possible; ii) finance irrigator retirement; or iii) await a better time for capital-transfer—have all been mentioned in the literature, and are all associated with a desire to keep farming or remain in local communities [27]. The important finding here is that much of the selling seems to be taking place during scarce water supply conditions in a forced rather than willing way due to debt, other financial or looming retirement reasons. The purchase of entitlements by anyone other than government buyers seems unjustifiable.

In relation to the third question the concept of diverse trade products such as derivatives or options have been portrayed elsewhere [4, 27, 28] as more efficient products that might appeal to irrigators and governments as a means of resolving supply issues in advance. This study supports the finding that alternative allocation products (e.g. supply option contracts and long-term lease arrangements) would appeal to irrigators. At present though, these findings are not generalisable to the larger population because the sample here is very small.

We intend to address this with a later quantitative survey. Also, the context is one of drought and scarce water supply; if the seasons returned to normal the attitudes and experiences reported here would no doubt be different.

5 Conclusion

Support for the hypothesis that irrigators would be reluctant to engage in providing environmental water via the sale of entitlements is mixed. While irrigators appear unwilling to sell, in many cases they are being forced to do so because of drought and financial pressure. Further, because the interviews only targeted irrigators currently active in trading—i.e. not those who have already exited or have never participated in any kind of water trading—it is difficult to state conclusive findings. There is, however, evidence to suggest that alternative allocation trade products could appeal to irrigators in the process of water recovery should the entitlement market falter for any reason. The research provides a valuable start to expand the theoretical understanding, but the findings need to be further tested to determine wider acceptance and policy value.

Acknowledgements

This research is part of a larger project funded by the Australian Research Council and supported by six industry partners: Murray-Darling Basin Authority, CSIRO Water for a Healthy Country Flagship Program, Department of Sustainability and Environment in Victoria, Goulburn-Murray Water, the New South Wales Office of Water and the University of Lethbridge in Canada.

References

[1] Crase, L., O'Keefe, S. & Dollery, B., Water buy-back in Australia: Political, technical and allocative challenges, *53rd Annual Australian Agricultural and Resource Economics Society Conference*, Cairns, Australia 2009.

[2] Ashton, D. & Oliver, M., An economic survey of irrigation farmers in the Murray-Darling Basin: Industry overview and region profiles, ABARE, 2008.

[3] Dixon, P., Rimmer, M. & Wittwer, G., Modelling the Australian government's buyback scheme with a dynamic multi-regional CGE model, Centre for Policy Studies, Monash University Melbourne, 2009.

[4] Bjornlund, H., Efficient water market mechanisms to cope with water scarcity, *International Journal of Water Resource Development,* **19**, pp. 553-67, 2003.

[5] Bjornlund, H. & Rossini, P., Fundamentals determining prices in the water market for water entitlements - An Australian case study, *International Journal of Water Resources Development*, **23(3)**, pp. 537-53, 2007.

[6] Wheeler, S., Bjornlund H., Shanahan M. & Zuo A., Price elasticity of water allocation demand in the Goulburn-Murray irrigation district, *Australian Journal of Agricultural and Resource Economics*, **52(1)**, pp. 37-55, 2008.

[7] Scoccimarro, M. & Collins, D., Natural resource buybacks and their use to secure environmental flows, Land & Water Australia, Canberra, 2006.

[8] Thoyer, S., How to reallocate water rights when environmental goals conflict with existing entitlements, *International Journal of Sustainable Development*, **9(2)**, pp. 122-37, 2006.

[9] Breckwoldt, R., Review of the 2007-08 water entitlement purchases: Final Report, Hyder Consulting, Melbourne, 2008.

[10] Bazeley, P., *Qualitative data analysis with NVivo*, 2nd ed., Sage Publications: London, 2007.

[11] Gladstone, J., Dupuis, S. & Wexler, E., Changes in family involvement following a relative's move to a long-term care facility, *Canadian Journal on Aging*, **25(1)**, pp. 93-106, 2006.

[12] Lincoln, Y.S. & Guba, E.G., *Naturalistic inquiry*, Sage Publications Inc.: Beverly Hills CA, 1985.

[13] Giorgi, A. & Giorgi, B., Phenomenology (Chapter 3). *Qualitative Psychology*, ed. J.A. Smith, Sage Publications: London, pp. 25-50, 2003.

[14] Patton, M.Q., *Qualitative evaluation and research methods*. 3rd ed., Sage Publications: Thousand Oaks CA, 2002.

[15] Hoepfl, M., Choosing qualitative research: A primer for technology education researchers, *Journal of Technology Education,* **9(1)**, pp.47-63, 1997.

[16] Russek, B.E. & Weinberg, S.L. Mixed methods in a study of implementation of technology-based materials in an elementary classroom, *Evaluation & Program Planning*, **16(2)**, pp. 131-42, 1993.

[17] Corbin, J. & Strauss, A., Grounded theory research: procedures, canons and evaluative criteria, *Qualitative Sociology*, **13(1)**, pp. 3-21, 1990.

[18] Glaser, B., The constant comparative method of qualitative analysis, Social Problems, **12(4)**, pp. 436-45, 1965.

[19] Aronson, J., A Pragmatic View of Thematic Analysis, *The Qualitative Report*, 2(1), pg. 4, 1994.

[20] Strauss, A. & Corbin, J., *Basics of qualitative research,* Sage Publications: Thousand Oaks CA, 1998.

[21] Bjornlund, H., Can water markets assist irrigators managing increased supply risk?: Some Australian experiences, *Water International*, 31(2), pp. 221-32, 2006.

[22] Bjornlund, H., Formal and informal water markets: Drivers of sustainable rural communities? *Water Resources Research*, **40** (W09S07), 2004.

[23] Wheeler, S., Bjornlund, H., Shanahan, M. & Zou, A., Modelling demand for water entitlements in the Goulburn-Murray irrigation district of Australia. *Water Down Under Conference*, Adelaide 2008.

[24] Wheeler, S., Bjornlund, H., Shanahan, M. & Zuo, A, Factors influencing water allocation and entitlement prices in the Goulburn area of Australia (Chapter 7). *Sustainable Irrigation: Management technologies, Technologies & Policies II,* eds. Y. Villacampa-Esteve, C.A. Brebbia & D. Prats Rico, WIT Press: Southampton, pp. 63-72, 2008.

[25] Bjornlund, H., The socio-economic structure of irrigation communities: water markets and the structural adjustment process, *Rural Society*, **12(2)**, pp. 123-47, 2002.

[26] Edwards, J., Cheers, B. & Bjornlund, H., Social, economic and community impacts of water markets in Australia's Murray-Darling Basin region, *International Journal of Interdisciplinary Social Sciences*, in press.

[27] Bjornlund, H., Markets for water allocations: Outcomes and impacts. *Water Down Under Conference*, Adelaide, 2008.

[28] Bjornlund, H., What is driving activities in water markets? *Water*, 30, pp. 30-6, 2003.

The drought, the irrigators, and their photographs: images from the inside

G. Kuehne[1,2] & H. Bjornlund[2,3]
[1]*CSIRO Sustainable Ecosystems, Australia*
[2]*CRMA, School of Commerce, University of South Australia, Australia*
[3]*University of Lethbridge, Canada*

Abstract

This paper describes a photo-elicitation study conducted with irrigators in Australia's Murray-Darling Basin during a time of severe drought. This method was chosen because of its ability to elicit the expression of more deeply felt beliefs and values than interviews by themselves. Using visual methods combined with personal interviews, the aim of the study was to uncover the influences on irrigators' decision-making and their expectations for the future. The analysis of the interview transcripts in conjunction with the analysis of the photographs shows that irrigators have become dispirited and often no longer see a future for themselves in the industry. It was also evident that non-profit-maximising values, such as lifestyle and the prospect of family succession, are still powerful influences on irrigators' behaviour, even during a time of such severe financial distress. This study suggests that the programs and policies aimed at assisting irrigators, even during this time when financial issues could have been assumed to have been their most pressing concern, could be better designed so that they provide a more compatible match with irrigators' values and attitudes. This would potentially lead to less social, environmental and economic distress for the individuals and communities of the irrigation regions.
Keywords: decision-making, irrigators, photo-elicitation.

1 Introduction

Drought in Australia, while a regular and expected occurrence, is not usually expected to be as persistent and as severe as that currently being experienced in

WIT Transactions on Ecology and the Environment, Vol 134, © 2010 WIT Press
www.witpress.com, ISSN 1743-3541 (on-line)
doi:10.2495/SI100211

the study area—the South Australian Riverland. The drought has resulted in the lowest recorded inflows into the Murray–Darling river system over an extended period of time [1]. This has had a profound effect on Riverland irrigators, who have been accustomed to receiving 100% of their irrigation water allocation. However, in 2006-07, due to the low inflows, this was reduced to 60%, then 32% for the following season and for 2008-09 to 18%. In addition to declining water allocations, Riverland irrigators have at the same time experienced declining returns from their produce, primarily winegrapes and citrus, and a weakened property market. These unfavourable conditions have depressed the morale of the region. In the 2006-07 season 30% of grape growers were considering exiting the industry [2], however, by 2007-08 this had risen to from 50% to 60% [3, 4]. Irrigators were forced to consider a range of measures for adapting to the commodity price slumps and the ongoing drought.

The aim of this research is to develop a better understanding of the influences on irrigators' decision-making under harsh and demoralising conditions—in this case a combination of severe drought and declining commodity prices. The study uses the decision of whether to sell water to the government for environmental purposes or not as the dependent variable. Research has been done on the influences on farmers' decision-making, but often without considering the context of the decision-making, and with a group of farmers coping with such trying conditions.

The measure examined in this study, the decision to sell water entitlements, is especially poignant given that there is little chance to continue operating a farming business in the semi-arid environment of the Riverland once the water has been sold.

2 Methods

This study uses a multi-method approach; quantitative and qualitative mail-out survey and photo-elicitation interviews. Mail-out surveys were posted to the Central Irrigation Trust irrigator mailing list on 12 November 2008 with 291 surveys returned (23% of those posted). The forty questions (eight qualitative) gathered information from respondents about their: 1) farming operation; 2) management response to the current water restrictions, and 3) personal information. Data were first entered into Microsoft Excel™ and then analysed by generating cross-tabulations and chi-square tests using the SPSS™ statistical package.

Irrigators who 1) returned mail-out surveys, 2) agreed to be interviewed, 3) were resident landowners, and 4) provided detailed responses to the open-ended questions of the survey, were invited to participate in the photo-elicitation interviews. All who were approached willingly agreed to be involved because they wanted to tell a story about how they were being affected by the low irrigation allocations and the low commodity prices. Digital cameras and the instructions for the photographic topics were sent to participants two weeks prior to the interviews. The ten photo-elicitation interviews were undertaken from 19 December through to 22 December 2008. The interviews were digitally recorded, fully transcribed, and then coded using Nvivo 8. Statements were coded if they illustrated the influences on irrigator's decision-making.

Photo-elicitation is a useful research method because it provides the respondents time to consider questions and decide which photographs to take to best reflect their perceptions or opinions rather than being confronted with questions in an interview requiring an immediate response. The photographs which are inserted into the interview setting become intermediaries in the interviewer-interviewee relationship; focussing discussion on the image rather than the interviewee and helping to overcome status differences. They can allow for reflection, prompt spontaneous interviewee comments, facilitate freer-flowing exchanges between interviewer and interviewee, and encourage reflexive attention to specific and broader issues [5] as well as uncovering meanings that might not be uncovered in face-to-face interviews [6]. The use of photographs also assists in bridging the divide, and reducing potential awkwardness that can exist between the researcher and the participant—two people who are rarely likely to share the same cultural backgrounds [6, 7].

In addition, because the participants are responsible for choosing what to capture in the photographs they are rebalancing the power relationship between the researcher and the interviewee [8]. This shift from a "researcher-centric construction of the social world" [9, p. 17] results in a more collaborative approach where the research is done 'with' participants rather than 'on' participants.

The choice of the photo-elicitation method was validated by an interviewee's comments:

> "I mean I know it's really good taking photos, you can sort of um talk about anything when you've got the photos. Yeah I'm actually thinking more about [the topic]...when I got this I wrote down things that I could take photographs of."

3 Results and discussion

In this section some of the results and discussion from the mail-out survey are presented, followed by quotes and discussion from the photo-elicitation interviews. These discussions are used to form a model (see Figure 1.) explaining the dominant role of succession (business continuity) in the decision to sell water to the government or not.

The qualitative and quantitative responses to the survey showed that those without an expectation of family succession have a greater interest in the sale of water because they do not require the water to develop a business to hand over to the next generation. Those who expect to have family succession are divided between, 1) whether they would sell or consider selling water and 2) having no interest in selling water.

Those unsure of family succession are strongly disinterested in selling water; however this could be expected to change when that uncertainty is resolved. Indicating a lack of optimism in the region, 60% of participants are unlikely or

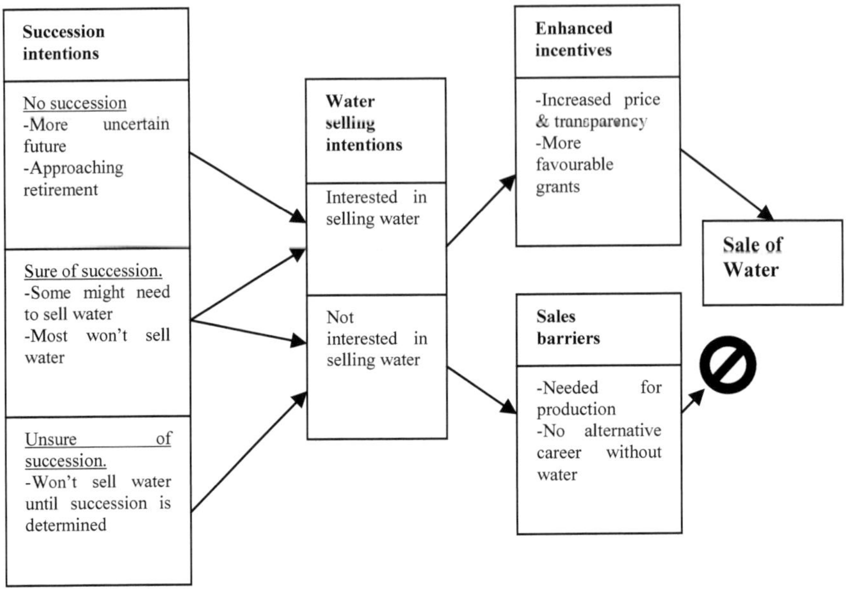

Figure 1: Influence of succession on the sale of water.

very unlikely to have family take over their farm business, 12% were unsure, and only 20% were likely or very likely to have family successors. This is half the proportion of irrigators expecting family continuity in 1998-99 [10].

Those who are certain or very certain of a long-term farming career are more unlikely to have considered selling their water. Although it may be that the farmers who have already sold their water, or who have committed themselves to selling their water at some point in the future, and those that see their sale of water as inevitable, would also anticipate a very uncertain career as a farmer.

Those farming because they can't sell their property are more likely to have considered selling their water. (Properties have sold, but at a substantial discount and although many were advertised for sale, few sales were occurring during the time of the research.)

Those who had, 1) no alternative career, or 2) liked the lifestyle, were unlikely to sell their water. Respondents who are uninterested in selling their water are not strongly influenced by the price they could receive for their water. This is likely to be because they have a productive use for their water that is of more value to them.

On the other hand some farmers are quite willing to sell especially if prices for water rose to levels higher than those currently on offer. They would also sell if access to the 'Small Block Irrigators' Exit Package' [11], the government's initiative to motivate irrigators to sell their water by providing a $150,000 grant, became less restrictive. An indication of the seriously weakened financial state of many of the irrigators in the Riverland is their desire to access the exit package regardless of the conditions which require the sale of all their water to the government, and a five year embargo on their property being used for irrigated agriculture.

Those in the fifty-one to sixty years of age group are more likely to have considered selling their water. The up to fifty years of age group are equally divided on wanting to sell their water or keep it, whereas the sixty-one years and older group are less likely to want to sell their water. It may be that the oldest group want to keep their water until the time that they retire from farming at which time they intend selling their water to provide a superannuation equivalent.

For the interview analysis, participants' photographs were combined with the parts of the transcript where the photographs were discussed. Due to the limitations of this paper only illustrative examples of the photographs are provided here. Other coded parts of the transcript are also presented where they are relevant to the discussion on influences on the decision to sell water or not.

The interviewee providing Figure 2 said:

> [This is]…the future … It's what I do it for.

This interviewee's work is being done for the family. But for others the current unfavourable conditions have influenced succession intentions.

> I think [my son] sees that he doesn't want to go through this, … we're on Exceptional Circumstances drought relief. That's what we're living on. …

Not everyone is defeated by the drought; some believe that the situation will eventually improve (Figure 3).

Figure 2: Family members are the focus of the future.

Figure 3: Optimism still exists.

> There's six acres out there ......it's a bit of optimism in tough times. So how can we explain that ... Faith basically ... and a fundamental belief in oneself. ...

At the time of the interviews farming offered a much less rewarding career than it had in the past. Many reported that their children had elected to find alternative and more rewarding careers than farming. Some farmers appeared to believe that they have been successful if their children did not become farmers.

> But still I don't regret it, I mean you know we had a family of six and the kids had a good upbringing and good country life and they've gone off and done better things.

The farmers seem to recognise that their children may not want to continue on with the business:

> ... They've seen what's happening, I mean they can see what it's done to us, you know you just struggle along.

The children of farmers who overcome the pessimism of others, and are still keen to take on the business, can be faced with difficult choices:

> He's starting to realise that to put it together the amount of money he needs ... won't give him a quality of life ... wanting to be a farmer can destroy you ... you really have to then commit everything of you to make it, you don't get the financial rewards of that until it's too late in life.

Family succession as a form of inter-generational change has been interrupted by current events.

> … we were trying to buy more land and, and the plan was to just do the expansion and then to gradually do the generational change. These water restrictions have stopped the ability to do that for us and I think for much of the district so the capacity to do the generational change is just about squeezed …

The drive to continue the farm business is strong for farmers with willing successors but it is less so for those without likely successors:

> I was seriously looking at expanding and buying more vineyards because I see the opportunity there but then I think … in the end why would I do it, because I don't need to concern myself about the whole family succession process.

A recurring comment is that growers feel trapped into continuing as farmers by the current circumstances.

> I can't see us getting out of this in the near future anyway 'cause you can't sell the property, nobody's interested in buying any more, unless you want to be trapped here you're going to take the best that's on offer and try and get out …

Some farmers have a strong desire to remain on their blocks and in their community. The people that would like to sell their water but remain living in the area often appear to be toward the end of their working life, and have an attachment to community that spans many years.

> I know people who want to stay because they've committed themselves … They've got nowhere else to go anyway so they're really committed to staying on.

Some farmers show the effects of the drought are not just financial but also emotional. It seems that some people who have made all the 'right' decisions are now the ones who are caught up in this crisis (Figure 4 shows the place referred to in the second of the following quotes).

> [W]e came up here for semi-retirement. I know it's hard but it develops character, perseverance and endurance, it really does. Within two years we'd worked out that we'd made a bloody great mistake buying it, it was the worst thing we ever did.

> That is my least favourite place, it's a fallen gum tree and it's fallen on the orange trees that are dead and it's impossible. … it's been sitting there for two years.

For many the present situation invokes feeling of powerlessness, being trapped, of stress and of hard work for little reward. The business environment has changed substantially, quickly and unexpectedly, so that someone who was once making a good living can now be faced with a business that is rapidly becoming unviable.

Figure 4: Powerlessness is overwhelming for some.

Figure 5: Hard decisions need to be made.

> ...it was a hard decision to make to - I mean you had to make it, everyone was saying yeah you've got to stop watering something... I've had years of good fruit on these trees...they're old. ... Old as me—fifty-five years.

This grower had serious difficulties in making the decision to let the trees die (see Figure 5) because of his long term relationship with them; it almost appeared as if he considered them as a part of his family and his life. The decisions he makes need to be made in light of the responsibility and the stewardship that he feels for these living things.

Generational change is likely to become a difficult issue for many of the horticultural and viticultural businesses of the Riverland.

> I'm forty-eight and now that's really scary going to meetings and all you see are old men, I see that the exit strategy is a perfect situation where if you can allow many of these people to retire from the industry …

Succession plans have needed to be adjusted, meaning that for many in the Riverland, trouble-free generational change has become less likely.

> When you take the water off a property … they're not just buying the water they're buying the whole business … And for a lot of guys around here it's their superannuation … It's their whole life savings. … Most growers around my age, I'm fifty-three, a lot of these guys are looking to retire … the idea was of course that you build your property up and if you're wealthy enough to do that you'd hand it on to a son or whatever, or if not you'd sell it to a neighbour.

Retirement expectations have now changed for many in the Riverland as the value of their property has reduced:

> It was our superannuation … we had two million dollars worth of property. We were going to somehow pass it over to our son. This was the option five years ago.

Water is valued by these growers because it allows an annual production of fruit, vegetables, nuts or wine grapes. But it is also water that determines the value of their property. The ability of irrigators to retire comfortably after a lifetime of hard work is made possible by the value contained in their property. Some intended that this retirement took place on their property, so that they remained living in the same house, but for others this was not so important.

4 Conclusions

This research has sought to elicit details about non-profit decision drivers during a period of extreme stress, hardship and uncertainty in the Riverland of South Australia. Responding to low commodity prices, an on-going drought and the Australian Government's moves to recover water from irrigators to be returned to the environment, many irrigators were considering the difficult decision of whether to sell water to the government or not. Because of the importance of water this is also a decision whether to leave farming or not. Many of the farmers who participated in this research are some years away from retirement, and have little prospect of finding fulfilling alternative employment should they exit the industry. Many are uncertain about plans for family succession or, if a family member plans to continue on in the business, whether the future rewards of farming will be worth the effort for them.

When considering the sale of water we found a confluence of non-profit and profit maximizing values influencing farmers' decisions. They include 1) whether the farmer wishes to continue farming in the future, 2) years they have left until retirement, 3) whether family succession has been arranged, 4) their future employability, 5) the eligibility criteria of the exit grant package, and 6) the price on offer for water.

Financial considerations do have a strong relationship with a decision to sell water, but non-profit decision drivers still continue to be important influences on farmers even during times of severe financial stress.

Acknowledgements

This research was funded by Rural Industries Research and Development Corporation, Dept. of Sustainability and Environment, Dept. of Primary Industries and Resources of SA and University of South Australia.

References

[1] Murray-Darling Basin Authority, *River Murray System Drought Update.* 2009, Murray-Darling Basin Authority: Canberra. p. 1-7.

[2] Chambers, M., *A survey of wine grape growing farms in the Murray Valley and Barossa regions, 2006-07*, in *ABARE research report 08.11*. 2008, Australian Government Department of Agriculture, Fisheries and Forestry: Canberra.

[3] Cummins, E., *SA growers prepare to quit struggling industry.* Grapegrowers and Vignerons, 2008. **October**: p. 4-5.

[4] Riverland Drought Taskforce, *Riverland drought taskforce regional submission* Productivity Commission's Government Drought Support Inquiry 2008: p. 1-13.

[5] Parker, L. *Photo-elicitation: an ethno-historical accounting and management research prospect. In 4th International Critical Management Studies*. 2005. University of Cambridge, Cambridge, UK.

[6] Clark-Ibanez, M., *Framing the social world with photo-elicitation interviews.* American Behavioral Scientist 2004. **47**(12): p. 1507-1527.

[7] Harper, D., *Talking about pictures: a case for photo elicitation.* Visual Studies, 2002. **17**(1): p. 13-26.

[8] Kolb, B., *Involving, sharing, analysing-potential of the participatory photo interview*. Qualitative Social Research, 2008. **9**(3).

[9] Prosser, J. and Loxley, A., *Introducing Visual Methods*, in *ESRC National Centre for Research Methods Review Paper*. 2008, National Centre for Research Methods. p. 65.

[10] Bjornlund, H., *Markets for water allocations: Outcomes and impacts, in Water Down Under*. 2008: Adelaide.

[11] Prime Minister of Australia. *New exit grant package to assist small-scale irrigators.* 2008 [cited 2009 27 January]; Available from: http://www.pm.gov.au/media/Release/2008/media_release_0487.cfm.

Author Index

Groundwater Characterization

M. CUNHA, University of Coimbra, Portugal, L.M. NUNES, University of Algarve, Portugal and L. RIBEIRO, Technical University of Lisbon, Portugal

This book addresses the theoretical background necessary to accomplish planning and management of groundwater systems, and presents up-to-date applications of the decision-aid techniques in this field.

Groundwater systems play an essential role in meeting the ever-increasing demand for water for various purposes. Proper design and management of such systems should therefore be a very important matter of concern, not only to ensure that water will be available in adequate quantity and quality to satisfy demands, but also to guarantee that this would be done in an optimal manner from a IWRM perspective. There are many different decisions to be taken: where to locate wells, how much water is to be pumped, remedial strategies to be adopted, water supply structures (especially pumping equipment and pipe networks) to be installed, monitoring networks to be defined, etc. These decisions must take many constraints into account, including drawdown limitations, flow gradients, and quality standards. Given the uncertainty characterizing groundwater flow and transport, risk issues have to be considered. Decision-aid techniques must be capable of handling simultaneously the various facets characterizing such problems (economic, social, technical, environmental, etc). Therefore detailed simulation models have to be incorporated into the decision models. The application of simulation-optimization methods to planning and managing groundwater systems has become an area of active research.

ISBN: 978-1-84564-134-4 eISBN: 978-1-84564-351-5
Forthcoming apx 350pp apx £115.00

Management of Scarce Water Resources

H. EL-NASER,
Overseas for Sustainable Development (OSD), Jordan

Arid and semi-arid countries have historically suffered from a plethora of complex water concerns due to climatic limitations. Research in these regions has long supported the necessity of delivering life-sustaining access and availability to water while being inherently linked to several inter-related factors, including technical and economic issues, human resources, health and private sector participation. Regional conflicts and controversial political agendas have also exercised a profound impact on the viability of co-operative water resource and management perspectives.
Dr Hazim El-Naser has applied his experiences as a water expert and goverment minister in Jordan by exploring a wide variety of water-related problems that the MENA countries face with respect to water resource management. This book focuses on his experiences, highlighting the complexity of these problems and advocating practical solutions, and uses recent case studies to provide a framework to assist the practitioner in resolving these problems.

Progress in Water Resources, Vol 14
ISBN: 978-1-84564-414-7 **eISBN: 978-1-84564-415-4**
2009 224pp £89.00

Water Pollution X

Edited by: ***A.M. MARINOV,*** *University Politehnica of Bucharest, Romania and*
C.A. BREBBIA, *Wessex Institute of Technology, UK*

Water pollution is essentially an interdisciplinary field involving scientists and professionals with a wide range of expertise. It also transcends national borders as the contamination of water resources is a problem of global concern.

The International Conference on Modelling, Monitoring and Prevention of Water Pollution, held regularly in different locations around the world, provides a forum for the presentation of the latest developments in the field and for their discussion. The papers in this volume present some of the latest results in this important field; work which is essential to understanding the nature of the problem and for proposing appropriate solutions, which may eventually provide the guidelines required to take steps towards the remediation or recovery of water resources.

This book contains papers presented at the Tenth International Conference and includes the following topics: Coastal Areas; Lakes and Rivers; Groundwater and Aquifer Issues; Environmental Monitoring and Sensing; Remediation; Experimental and Laboratory Work; Mathematical Modelling; Oil Spills; Wastewater Treatment and Management; Low Cost Technologies; Agricultural Contamination; Pollution Prevention; Emerging Technologies for Water Treatment; Novel Techniques for Water Treatment; Pharmaceutical and Pesticides; Bioaccumulation; Micropollutant Prevention in Drinking Water.

WIT Transactions on Ecology and the Environment, Vol 135
ISBN: 978-1-84564-448-2 **eISBN: 978-1-84564-449-9**
2010 384pp apx £142.00